Survival, Disaster and Tactical Medicine

By Chris Breen, RGN, Paramedic

Copyright © 2011 by Chris Breen

Cover design by David C & Chris Breen

Cover art © Ermess, Joyfull, Enrico Bragante & Porojnicu from Dreamtime.com

Book design by Chris Breen

All rights reserved.

No part of this book may be reproduced in any form or by any electronic or mechanical means including information storage and retrieval systems, without permission in writing from the author. The only exception is by a reviewer, who may quote short excerpts in a review.

Chris Breen

Printed in the United Kingdom

First Printing: May 2011

Version 1.4

ISBN: 978-1-4477-1225-1

Dedication

Many people have helped in this project, I would like to thank Caroline Jackson for all her support, Joe Jackson, Di, Nigel, Emma and Angela for help with the photographs.

This Book is inspired by and dedicated to my friends at the Ludlow Survival Group. Particular thanks to Dave and Maureen D, Tim & Gill Anderson, Jason F, David C, Hugh & Fi, Andy and Nix W and my colleagues at EMAS.

Updates

If you were like to receive information on updates, corrections and new editions. Or if you have any suggestions for new content. Please email me at;

Stone.temple@ntlworld.com

INTRODUCTION 7

CHAPTER 1 EXAMINATION OF PATIENTS 11

CHAPTER 2 ENVIRONMENTAL PROBLEMS 55

CHAPTER 3 BITES AND STINGS 62

CHAPTER 4 DEALING WITH PARASITES 68

CHAPTER 5 FOOT CARE 73

CHAPTER 6 TRAUMA 75

CHAPTER 7 JOINT PROBLEMS 184

CHAPTER 8 ALLERGIC REACTION 187

CHAPTER 9 RESPIRATORY ILLNESSES 191

CHAPTER 10 ABDOMINAL ILLNESSES 202

CHAPTER 11 NEUROLOGICAL PROBLEMS .. 210

CHAPTER 12 CIRCULATORY PROBLEMS 215

CHAPTER 13 DIABETES 224

CHAPTER 14 INFECTIOUS DISEASES 231

CHAPTER 15 POISONING.................................. 240

CHAPTER 16 SHOCK... 243

CHAPTER 17 EAR, NOSE AND THROAT (ENT) 254

CHAPTER 18 EYE PROBLEMS 260

CHAPTER 19 SKIN CONDITIONS 266

CHAPTER 20 MINOR MEDICAL PROBLEMS. 267

CHAPTER 21 MEDICATION 269

CHAPTER 22 VACCINES (IMMUNISATION) 295

CHAPTER 23 ROUTES OF DRUG ADMINISTRATION 298

CHAPTER 24 CLINICAL SKILLS 305

CHAPTER 25 MEDICAL KITS & SUPPLIES ... 365

CHAPTER 26 EMERGENCY DENTISTRY 381

CHAPTER 27 SEXUALLY TRANSMITTED DISEASES (STD) 392

CHAPTER 28 REPRODUCTIVE PROBLEMS .. 394

CHAPTER 29 LONG TERM CARE 406

CHAPTER 30 TACTICAL CONSIDERATIONS 409

CHAPTER 31 TRIAGE 411

APPENDIX 1 GCS 414

APPENDIX 2 MEDICAL TERMINOLOGY 416

INDEX 418

Introduction

The author is a Registered Nurse who served with the RAMC, a Paramedic and Clinical Tutor with additional qualifications in Trauma and Remote Medicine. He has had a long term interest in Survival medicine and is the Medical Advisor for a Survival group and runs courses in Survival Medicine.

The contents of this book are derived from a number of articles which have been published online and the syllabus of the Wilderness Medicine Course he runs.

The aim of this work is to provide the lay person with the knowledge and skills to deal with a variety of medical conditions and traumatic injuries usually dealt with by Health Care Professionals.

Equipment, medical supplies and initial first aid treatment is covered, if the injury or illness would benefit from more advanced measures then these are detailed as well as any skills needed to carry them out. We will also look at aftercare and the limitations of care without the benefit of a modern health service.

If access to emergency services or normal health care is obtainable this should always be done, the advanced advice and techniques in this book should only be practiced by lay people if no other help is available and would not be for some time.

Scenarios where this could apply is in a Survival situation, where the time normally taken to be rescued may be delayed or when involved in major disaster where normal health services are unable to cope or have ceased to function.

Life is uncertain, people who are involved in survival situations, disasters and accidents are often unprepared both in knowledge and supplies. This book is intended to give you a grounding in medical care and encourage you to prepare supplies suitable to your level of knowledge and likely situations you may encounter.

It is hoped it will be useful to those who partake in extreme and wilderness sports and activities as well as those who are interested in survivalism and preparedness.

However reading a book is never a substitute for proper training and experience. Anyone seriously interested in using this knowledge should seek additional training with a reputable training organisation to the desired skill level. The author takes no responsibility for the use of the information contained herein by non Health Care Professionals.

Typical Injuries from sports

Mountain Bikes

Cyclists particularly mountain bike and dirt bikers, have a high incident of injuries usually this is just bruises and abrasions although fractured wrists, collar bones and lower leg fractures are not uncommon. A study published by Emily R. Dodwell et al, titled "Spinal Column and Spinal Cord Injuries in Mountain Bikers: A 13-Year Review" in Am J Sports Med, May 2010 in America also showed that between 1995 and 2007, 102 men and 5 women received spinal injuries. The average age being 32, around three quarters of these damaged there cervical spine where it run through the neck. A third of those injured required intensive care, two thirds required surgery and two were so badly injured they still require portable ventilators.

Runners

Not surprisingly the majority of runners injuries are sustained to the lower limbs and include the hip, leg, knee, ankle and foot. Types are numerous and can involve stress fractures, dislocations as well as damage to muscles, tendons, joints, cartilage and bone. Unfortunately runners are also involved in road accidents and falls that can produce a variety of injuries. Due to the nature of the sport they are often wearing minimum clothing and no protective equipment that increases the potential for injuries.

Skiing

There has been a lot of jokes around skiing injuries and accidents fortunately the incidence of injuries in skiing has almost halved in the last 15 years mainly due to better training, safer practice and the use of protective equipment. Although falls can cause a variety of problems some of the most common injuries are to the knee mainly the anterior cruciate ligament (ACL), Medial Collateral Ligament and crescent-shaped cartilage menisci which supports the knee.

Horse Riding

Most horse riding injuries are sustained during jumping or cross country although some are sustained in gentler pursuits. Horses are big, fast and often unpredictable they can weigh up to 1,500 pounds, travel as fast as 30 mph, and stand as tall as 3 meters high. Any fall from or collision with a horse is going to hurt. One study showed and for every 350 hours you spend on a horse you are likely to have an accident compare this to motorbikes which average an accident ever 7000 hours of riding. Most riding accident are caused by falling from or being thrown from the horse. Soft tissue injuries are common as well are fractures to wrist, elbow and collar bone where people instinctively put out there hands to break the fall. The use of helmets are significantly reduced the incident of head injuries and using kevlar body protectors that of rib fractures and internal injuries. Unfortunately due to the speed and size of the horse long bone fractures and spinal injuries are still sustained, as well as crush injuries if the rider gets trapped by or rolled on by the horse.

Climbing

Apart from the obvious danger of falls and scrapes from banging on the rock face, climbing can produce some unique injuries. As a sport it requires more upper body strength and it's the upper body that suffers. Overuse of fingers on holds can cause ligament strains, flexor tendon strains, tendonitis, ruptures and stress fractures to knuckles. Other common injuries are to wrist, elbows and shoulders but like most sports most joints can be affected.

Environmental factors

In any wilderness setting the climate and environment can cause problems in themselves with dangers ranging from heat stroke to frostbite, insect stings and bites and altitude related illnesses.

Chapter 1 Examination of Patients

In order to diagnose an illness or ascertain the extent of an injury a full patient examination should be made. The full sequence below is primarily aimed at a seriously injured trauma patient. But elements of the assessment can be applied to any patient.

Patient Position

Always consider the most appropriate patient position for examination. If the patient may have suffered serious traumatic injury before examination lay casualty flat on their back. If collapsed and still unconscious but without injury consider placing in the recovery position. If conscious sit or lay as appropriate. Anyone with breathing problems should be sat up or propped up if lying. If shock is suspected then lay down and raise legs, unless they are injured. If shocked and breathless, support head and shoulders but still raise legs if practical.

Getting started

Before you touch or speak to the patient take a mental step back and look at the patient called an 'end of bed' assessment. This initial survey can tell a lot about them. How do they appear? Are they pale, sweaty, shaking, do they appear anxious, how are they sitting are they holding an injured or tender part of their body?

Do they appear to be struggling to breath?

Can you hear any wheeze or rattle when they breath?

Is there anything around that gives clues to the situation?

Don't overlook the obvious ask the patient what's wrong, they might have experienced it before and know what the problem is and what needs to be done to resolve it.

In patient assessment the aim is for early identification of any potentially life threatening problems these are sometimes referred to as 'Red Flag indicators'. The extent of an illness or injury can be judged on a sliding scale usually termed as between being 'Well' and 'Very' or 'Big Sick'. It is usually easier to say if a patient is at either end of the scale but more difficult to place where a patient is along the length of the scale.

First Aid students are taught the DRS ABC approach to casualty examination this will identify life threatening problems with the casualty. DRS ABC stands for;

- Danger
- Response
- Shout for or get Help
- Airway
- Breathing
- Circulation

If the patient is conscious and talking then it is safe to assume that they have a clear airway and an adequate breathing rate and pulse. But these observations still might not be normal due to illness and injury so still need to be assessed.

If a patient has a problem with their airway that needs attention before moving on. Without a patent airway both breathing and circulation are quickly compromised. Similarly a problem with breathing needs to be resolved before circulation is assessed.

The main aims of First Aid in this instance can be described as follows;

To (P)reserve the casualty's condition

To (P)revent deterioration in the casualty's condition

To (P)romote Recovery

There are collectively known as the three P's

Two basic type of examination are used firstly one for Trauma and unconscious patients who have had an accident or have been injured and secondly those who have a medical problem. Many parts are common to the two types of assessment. If performing a medical each body system needs to be assessed. In practice however a medical assessment tends to be more focused on the presenting problem.

A Health Care Professional when rendering Medical aid will expand the DRS-ABC sequence to ABCDEFGHI, doing a full assessment of the casualty. They then use advanced interventions to preserve life, prevent deterioration and promote the recovery of the casualty.

Scene Assessment

D- Danger:

Danger can come in many forms. It may be environmental, caused by adverse weather condition or the remoteness of a location. It may be manmade such as with traffic on a road or electricity in a building. Danger may come from a confused patient or an intoxicated or aggressive bystander. Slippery or uneven ground can pose a danger to the rescuer, particularly if they rush to help an injured friend then become a casualty themselves. Also before kneeling next to the casualty check the ground next to them for sharp objects and bodily fluids.

As you approach the scene of an incident, you must be constantly aware of (and be continuously reassessing for) potential dangers, and be aware of the casualty and the area around them.

R- Response

After looking for any potential dangers before approaching a casualty, the first Vital Sign we measure is the Patient Level of Response or Level of Consciousness (LOC). This can be measured simply as Conscious or Unconscious or using an extended scale such as the Glasgow Coma Scale (GCS) which is difficult to remember. A good compromise is to use the AVPU Scale.

A = Awake/ Alert

Alert and oriented to: Time, Date, Place and recent events

V= Responds to voice

Responds appropriately

Confused

Makes incomprehensible sounds (Grunts, groans, etc.)

P= Pain

Response in some way to a pain stimulus

U= Unresponsive

No Response

Anyone who is not Awake & Alert should have their level of consciousness (LOC) monitored and constantly reassessed.

Use of Glasgow Coma Scale should be used to guide the management of head injured patients. The Glasgow Coma Scale is difficult to apply to children under 5 years of age. Although modifications exist, great care needs to be taken with its interpretation. (See Appendix 1)

S- Shout for help.
This applies to normal situations where someone may be close by to assist you, fetch first aid or medical equipment or call for an ambulance.

Primary Assessment

A - Airway:

Secure the airway while taking precautions to stabilise the cervical spine (see Spinal Immobilisation), if indicated.

If the casualty is conscious, they will have assumed a position where they can breathe comfortably. Any further interventions must not impair their capacity to breathe.

If the casualty is unconscious, log roll them onto their back and examine their airway. 'Look', 'Listen' and 'Feel' for movement of air.

By placing your left ear near the casualty's mouth you will feel the presence of the breath and will hear any sounds that the breathing produces. If air movement is partially obstructed then the amount of air you feel will be decreased and breathing noisy. Noise on breathing in *(Inspiration)* is indicative of a blockage in the upper respiratory system, whereas noise on breathing out *(Expiration)* indicates a lower blockage. The tongue partially blocking the windpipe may cause a 'snoring' sound or a 'gurgling' sound would suggest liquid or semi-solid matter such as vomit in the windpipe.

The simplest method of opening the airway is by tilting the head and lifting the chin. This is achieved by pushing the forehead backwards with your left hand whilst supporting the back of the neck. Then place two fingers of your right hand under the tip of the jaw and lift the chin as this will move the tongue from the back of the throat and help clear the airway.

Briefly check mouth for obstructions either from the tongue or the presence of blood, vomit, oedema, loose teeth, dentures or other foreign matter. If you see any debris, sweep your fingers in the casualty's mouth to remove it. Do not sweep blindly as this may increase the obstruction.

If you have suction equipment or airway adjuncts they can now be used. However if the patent is maintaining their own airway it is often better to avoid unnecessary interventions as these can trigger the gag reflex and cause the patient to choke. (See Clinical Skills Chapter)

After any airway has been inserted check if it works by repeating the 'Look', 'Listen', 'Feel' procedure to detect breath sounds and movement.

While examining the airway, check for any smells on the breath such as Alcohol, Cannabis, Pear drops which indicate the presence of Ketones in patients with High blood sugar (hyperglycaemia) , Solvents etc.

A patient who is unable to talk or is hoarse may have swelling, damage or a blockage which could compromise the airway

B - Breathing:

Once an airway is established look down the body. If the windpipe is completely blocked but the casualty is still making a respiratory effort then you may still feel and see chest movements, so the presence of breath must be verified. If the blockage is in one of the branches of the windpipe leading into the lungs then chest movement may be uneven.

Observe for a maximum of 10 seconds in that period you should see and or feel at least two breaths. If breathing is inadequate CPR must be started. This can be achieved by mouth-to-mouth ventilation with or without an airway adjunct or with a bag, valve, mask device (BVM)

RESPIRATION RATE

Is the breathing:

Normal

Normal breathing is regular, un-laboured, quiet and off moderate depth.

Deep

Excessively deep breathing

Shallow

Very small breaths

Laboured

Indicators of laboured breathing are if the patient is leaning forward called the tripod Position, nasal flaring, retractions which are the inward movement of the muscles between and below the ribs as a result of reduced pressure in the chest, accessory muscles use which is the use of the shoulder, neck and other muscles to expand the chest cavity and allow more airflow.

Normal respiration for Adults is *12 - 20 breaths/minute*

Abnormal 10-12 and 20-30 at rest

Serious <10 or >30

Normal Child range varies with age:

30-40 Resps/min between birth – 1 year Old

20-30 Resps/min between 2 – 4 years Old

15-20 Resps/min between 6 – 12 years Old

12-16 Resps/min at 14 Years Old

Remember: An increased respiration of an injured patient at rest may be the first sign of developing shock.

C - Circulation:

Determine pulse rate and blood pressure. In conditions with serious blood loss, fluid replacement is indicated. Only give fluids in the absence of a radial pulse and control the volume given until a pulse is restored.

Control haemorrhaging with direct or indirect pressure or application of a tourniquet. If circulation is not present begin CPR or Defibrillation (See Clinical skills)

Capillary Refill Time (CRT)

The CRT is the time it takes blood to return to an area after it has become blanched. CRT can either be measured peripherally on a nail bed, hand or limb. It can also be measured centrally on the chest or forehead

Press on the area for five seconds; it will go pale, then release, if the skin takes more than two seconds to re-colour it indicates reduced circulation. This is an unreliable measure if the patient is cold or already has circulatory problems.

A deficit in the peripherally circulation indicates a circulation problem. A deficit in the central circulation is a serious sign and may be due to shock.

Tissue colour is a good indicator of the state of circulation if you check the inside of the mouth and the lips are pale then the problem is peripheral if the tongue is pale then the problem is central.

Pulse

A pulse needs to be obtained this is usually taken at the radial site in the wrist but can be taken at the neck (Carotid Pulse) or anywhere an artery crosses over a bone and is close to the skin's surface. Other sites are the Groin (femoral), Upper arm, between biceps and humerus (brachial pulse, Head (temporal), Top of foot (Dorsalis Pedis), back of knee (popliteal). When feeling for a pulse use two or three fingers as the increased surface area will make location easier.

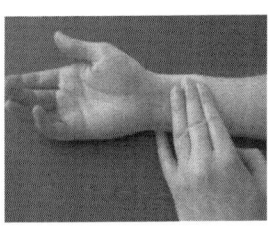  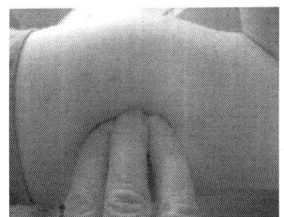

When assessing the Pulse the following should be taken into account:

RATE Either count for 15 Seconds and multiply by 4 or for a full minute.

Average Adult	60-80 Beats/minute
14 Years	80-100 Beats/minute
6 Years - 12 Years	80-120 Beats/minute
2 Years - 4 Years	95-140 Beats/minute
New Born - 1 Year	110-160 Beats/minute

In Adults <60 is Bradycardia (slow pulse) >100 is Tachycardia (fast pulse)

RHYTHM either:

Regular

Regularly Irregular (With Extra Regular Beats)

Irregularly Irregular (with no discernable pattern, most often a rhythm called Atrial Fibrillation or AF)

QUALITY

Whether Normal, strong and bounding, or weak and thready.

LOCATION

Location is important for three reasons:

Firstly by checking the pulse at the wrist (radial) and the pulse in the neck (carotid) you can roughly guess a blood pressure.

Radial pulse present = B/P $\geq$ 80 systolic

Femoral pulse Present = *B/P $\geq$ 70 systolic*

Carotid pulse present = *B/P $\geq$ 60 systolic*

Secondly, having unequal pulse in two arms may indicate a cardiac problem.

Thirdly, lack of pulses in a limb could indicate damage to the vessels from disease, direct or indirect trauma.

D – Deficits (Neurological): Level of Consciousness (LOC)

Assess the level of consciousness (LOC) by using AVPU (See Above) or GCS (see Appendix 1),

Check Pupils are Equal, Reactive to Light and Accommodating (PERLA).

Application of Glasgow Coma Score should be used to guide the management of head injured patients. However the Glasgow Coma Scale (See Appendix 1) is difficult to apply to a child under 5 years. Although modifications exist, great care needs to be taken with its interpretation.

Now move on to the secondary assessment

Secondary Assessment

E – Exposure / Search

In an unconscious or trauma patient remove clothing (Exposure) and look for additional wounds and any other unseen injuries (bleeding, bruising, burns or deformity). After an area is checked replace clothes or cover with a blanket to prevent heat loss. Check for medical alert bracelets, pendants or cards. Check if they are carrying any medication, a basic knowledge of common prescription medication will give you a good idea of a person's medical history providing of course the medication is theirs.

F – Fahrenheit

After exposing the casualty to check for further injury, care must be taken to preserve body heat (Fahrenheit) whilst performing any procedures required. If you have already started giving IV Fluids this will chill the body unless they have been pre-warmed.

G – Get a Set of Base Vital Signs

When assessing patients it is important to obtain a set of observations, sometimes known as vital signs, this is required for four main reasons;

> To aid identification of the underlying problem
> To gauge the severity of injury or illness
> To monitor the progress of the patient's condition
> To assess the effectiveness of treatment on the patient's condition

Vital signs include;
> Level of Consciousness (see above)
> Pulse (see above)
> Respiration Rate (see above)
> Blood Pressure (see under clinical skills)
> Temperature (see below)
> Capillary Refill Time (CRT) (see above)
> Skin Colour and Turgor (see below)
> Blood Glucose (see under clinical skills)
> Oxygen saturations (see under clinical skills)

TEMPERATURE

Normal Temperature range is 36.5 – 37.5 degrees Celsius.

Temperatures can be recorded with glass mercury, disposable paper or digital oral or tympanic (Ear) thermometer. Fever strips are also available to place on the forehead.

Four locations for placement are oral (under tongue), axilla (under armpit), tympanic (Ear) or rectal. Non digital, fever strips are also available which can be placed on skin and temp dot disposable thermometers which are used in place of glass mercury ones. thermometers should be left in place for 3 minutes. Axilla temperatures are generally one degree lower than oral ones.

SKIN COLOUR

Variation in skin colour can tell you alot;

PINK = Normal

PALE, WHITE, GRAY= Can indicate *Shock*

FLUSHED (RED) = Carbon Monoxide, High Blood pressure, Fever

BLUE = Hypoxia (Decreased Oxygen)

YELLOW / JAUNDICE = Liver Injury / Failure, Hepatitis, Cirrhosis

Texture: Clammy, Wet or Dry
Temperature: Cold, Warm or Hot

Skin Turgor

Turgor or tenting is a measure of the elasticity of the skin. It can become reduced if the patient is dehydrated by around 10% or more. Dehydration is often caused by severe diarrhoea and or vomiting or la decreased fluid intake. Infants and the elderly are most at risk especially those with a fever.

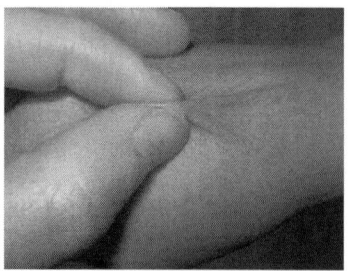

To assess lightly pinch some skin on the back of the hand, forearm or abdomen. Hold for a 5 seconds then release. Normally the skin would snap back to its normal position. If however the patient is dehydrated the skin returns slowly to normal or remains peaked.

H - History

If the patient is conscious ask them what happened, the events, signs and symptoms that led to their current condition. If not and If there were any witnesses to the incident or accident obtain as much information as possible from them. If they know the casualty asks about past incidents, medical history & any current medication or drug allergies. See SAMPLE OPQRST under medical patient below.

Then perform a Head to Toe examination

H - Head-to-Toe Examination

Head to toe examinations are most commonly performed on patients who have been injured in a way that might cause multiple problems, such as falls and road accidents.

Each area of the body should be examined using the DCAPBTLS system looking for the presence of;

(D)eformity, (C)ontusion, (A)brasion,
(P)uncture/Penetrating Injury, (B)urns, (T)enderness,
(L)aceration & (S)welling

The areas are examined in order of importance, the area's most likely to kill or seriously disable the patient are examined first.

Additionally different areas have specific methods of examination;

Head

Examine the scalp for wounds. Look for blood, fluid or vomit in the mouth, blood or fluid in the ears. Observe for bruising behind the ear ("Battle sign") and bruising around the eyes ("racoon eyes") which may indicate a fracture of the base of the skull.

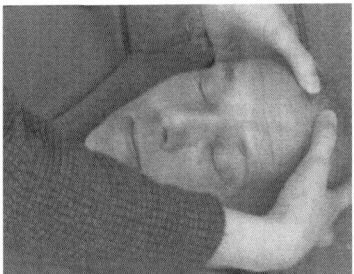

Yellow tinged blood or fluid coming from the nose or ears indicates the presence of cerebral spinal fluid CSF. This fluid encases the brain and its presence in blood indicates a skull fracture.

Whist examining the head recheck the pupil reactions. Look in the eyes and check for redness or a puffy appearance, is they blood or pus present? Are the eyes yellow suggesting the patient is jaundiced.

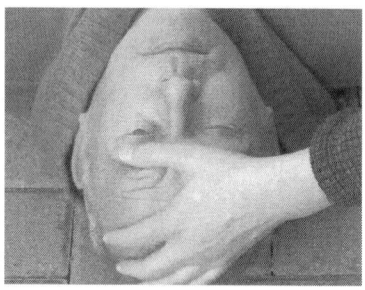

Wounds to the head, face, mouth or neck may suggest possible cervical spine injury.

Check central capillary refill on forehead.

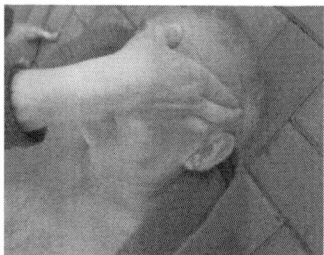

Airway compromise from facial injuries is potentially lethal due to haemorrhage, swelling and debris. Immediate stabilisation of the airway is imperative. Airway patency should be re-evaluated throughout care and transport.

Neck

Distended neck veins may result from a tension pneumothorax (Punctured Lung) or cardiac tamponade (Punctured Heart). Tracheal deviation may indicate a tension pneumothorax although this is a late sign and the patient would be seriously ill at this stage.

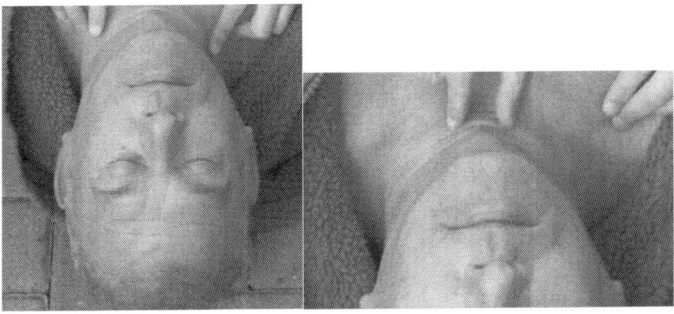

Crumpling cellophane sensation under the skin of the neck may indicate a pneumothorax with subcutaneous emphysema (Air in tissue).

Neck wounds require aggressive airway management due to the potential for rapid deterioration. Intubation should be attempted immediately in an unconscious patient if increasing neck swelling may compromise patient's airway.

Torso

Trunk wounds may consist of damage either to the chest or abdominal cavity.

Chest

Examine the chest using DCAPBTLS see above. Observe for equal and symmetrical chest rise. Place thumbs on(breast bone (sternum) and fan fingers out along each side of ribs. Using firm pressure to check for deformities in the chest wall by moving hands down sternum and ribs. When the patient exhales place thumbs tip to tip on their sternum, now as they inhales your thumbs should move apart equally. This will give you an indication of the symmetry and depth of breathing.

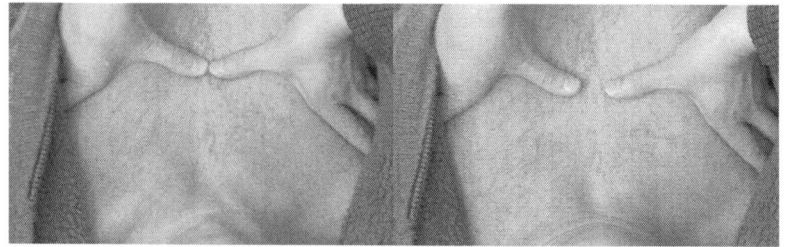

Using a stethoscope listen to the chest for breath sounds (Auscultation). Start at the Front (Anterior) of Chest and listen from side to side then move from top to bottom avoid the areas covered by bones

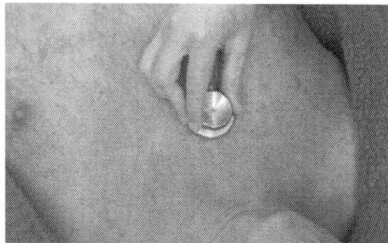

Compare one side to the other looking for differences. Note the location and quality of the sounds you hear.

In a trauma assessment you are primarily concerned with the presence or absence, rate and depth of breathing and noises that indicate a blockage to the airways. You may also hear other sounds these are detailed under Respiratory assessment below.

Abdomen

Examine the chest using DCAPBTLS see above.

Feel the abdomen to check for tenderness, distension or rigidity which can indicate damage to organs and internal bleeding. For a more detailed abdominal examination see medical assessment below.

Pelvis

If the pelvis is fractured poor handling can cause significant blood loss. See trauma chapter for more details. Look for signs of incontinence this can indicate spinal damage or seizure activity.

Legs and Arms

Examine the legs first as these contain the larger bones and blood vessels from which patients can lose the most blood. Move on to collar bones, shoulders, arms and hands. When you reach the wrist check for a radial pulse and check peripheral capillary refill at fingers

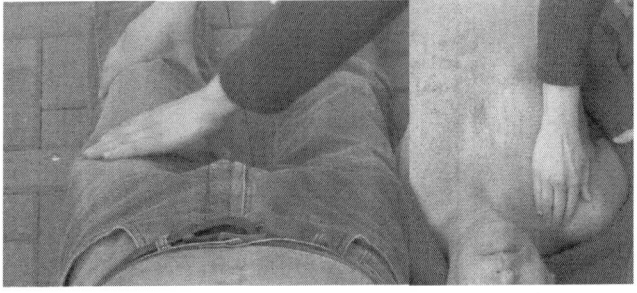

Evaluate extremities using DCAPBTLS.

Then check for Circulation, Sensation and Movement (CSM):

(C)irculation

Is the skin a normal 'Pink' Colour or are extremities pale or cyanosed? Can pulses be felt away (distal) from the site of the injury?

(S)ensation

Can the casualty feel you touching them, is the sensation the same on both sides?

(M)ovement

Can the casualty move limbs, fingers and toes? Does pain restrict movement?

I – Invert ("it's not over until they are over")

Turn the casualty over using a log roll to examine their back. Feel along the spine for irregularities and tenderness.

Medical Assessments

Some common sense needs to be applied if the above is applied to a medically ill patient then you need not check for bruises, cuts etc unless the patient also has a history of falls or other trauma. But assessment of Airway, Breathing and circulation is important for all patients as well as an examination of the chest and abdomen also obtaining observations such as Blood Pressure, Pulse, temperature etc

The art of questioning and obtaining a good history is the key to patient assessment and takes practice. An experienced practitioner will often come quickly to a working diagnosis after which the questioning will aim to confirm or exclude that diagnosis, They will often also have one or more differential diagnosis's to work with.

Phrasing Questions

Always ask open questions or seek clarification rather than used closed questioning technique. It is a well documented that when asking closed questions patients often agree with you even if they are not experiencing what you ask them about. This is because patients often feel that the questioner knows more about the condition than they do and if asked if they suffer a particular symptom feel they should have it so agree to please the questioner. It is better to ask "What sort of pain is it" rather than "Is it a crushing pain". After you have finished asking questions summarise the information and ask if your impression is correct and do they wish to change or expand on anything.

SAMPLE History

A SAMPLE history is a structured way of recording a patient's history.

- Signs and Symptoms
- Allergies
- Medications
- Past Medical History
- Last Eaten
- Events leading up to current situation

Gathering this information, even in a first aid situation is useful to pass on to the emergency services in the event the patient becomes unconsciousness before they arrive as it gives them clues to the patient's condition and therefore treatment options.

Signs and Symptoms

A Sign is anything you can see, hear, smell or feel that is pertinent to the patient. Typical signs include bleeding, pale skin, deformity of a bone, noisy breathing, smell of alcohol or crepitus etc. A Symptom is anything the patient feels; hot, cold, tired, thirsty, nauseous etc

Allergies

It is important to note any allergies or drug sensitivities, as this may alter the treatment a patient receives. Things such as Aspirin, Ibuprofen, Penicillin or any other drug or reactions to substances such as latex need to be ascertained.

Medication

A list of the patient's medication will give some clues to their medical history. It never ceases to amaze me how many people I go to who claim to have very little wrong with them but are taking over a dozen different drugs. These people are either ignorant of why the medication was prescribed to them, trusting in what the doctors tell them, or are in denial about their multiple medical problems.

Past Medical History

A list of pertinent points of interest about the person's Past medical history is important to record, the fact that a person has high blood pressure or diabetes is important, the fact they broke their wrist 20 years is less so.

Last Eaten or Drank

The recording of when they last ate or drank used to be done in case they needed an operation, in modern times they would have the operation regardless. It is more important with a diabetic patient or someone who is dehydrated.

Events

The sequence of events that led up to a particular incident can give clues to the cause. For example, if a person is found to be unconscious it's important to know what happened prior to the event. Did they hit their head or fall, did they complain of head or chest pain, have they been drinking and are intoxicated.

If the patient has pain them this can be assessed using the OPQRST formula.

(O)nset

When did the pain start, what were you doing or had been doing before it started.

(P) Provoke, Palliate or Prevent

Does anything bring the pain on (Provoke), Lessen (Palliate) on or Prevent it. Pain that is more intense when taking a deep breath is more likely to be from a problem with the lungs or the muscles between the ribs (Intercostals). But a patient with a constant central chest pain is more likely to have a heart problem.

(Q)uality

What's the pain like is it Sharp, dull, crushing, tight. Be careful with this as the patients idea of what the pain feels like may not fit with a textbook answer. A person may describe both the discomfort caused by a heart attack and asthma as a heavy pain. But others could separately describe them as crushing and tight.

(R)adiation

Does the pain go anywhere else, both chest and abdominal pain can often spread through to the back, arms, neck and shoulders.

(S)everity

How bad is the pain, is it at a constant level or does it come and go. A good measure is using a Pain score, ask the patient to set a number against the pain 'Where zero is no pain and ten is the worst pain they have experienced" again this can be subjective but is a good measure to see if treatment is effective. Score the pain at the beginning and after each intervention that could relieve it.

(T)ime
How long and how frequent is the pain.

Body Systems
Once you have a clear idea of the patients symptoms and have identified the signs they are exhibiting you are on the first step to making a diagnosis from which you can plan your treatment. The body is divided into systems a disease process will affect one or more of these systems. There are 11 body systems these can be divided into 7 groups as below;

- Cardiovascular System
- Respiratory System
- Nervous System
- Digestive System
- Genitourinary System (Reproductive & Urinary)
- Endocrine, Lymphatic and Skin
- Musculoskeletal System

Each System has key symptoms that would indicate a problem with that system. Some symptoms such as nausea are present in a variety of illnesses across a range of systems. These can be used to support a diagnosis only. Temperature usually indicates an infection this is most commonly in the respiratory problem, In the absence of other respiratory symptoms, think of other systems such as an abdominal problem.

Cardiovascular System
This System consists of the Heart and the blood vessels. Its purpose is to transport oxygen, water and nutrients to the cells and remove waste from them.

Symptoms
Cardiac Chest pain
Tiredness and breathlessness on exertion
Fainting and dizziness

Ankle swelling in the absence of infection and injury
Breathlessness when lying down
Awaking at breathless at night
Palpitations

Respiratory System
This system comprises the airways from the nose and mouth through the pharynx, larynx, trachea, bronchi to the lungs. It is supported by thoracic cage (Ribs, spine and sternum and the circulation to the lungs (Pulmonary). Its purpose is to provide ventilation of air to the lungs, Exchange gases between air and blood and perfuse the lungs with Blood

Symptoms
Chest pain worsened by breathing (Pleuritic Pain)
Acute or Chronic breathlessness
Hoarseness
Cough, sputum, Blood in Sputum (haemoptysis)
Wheeze

Nervous System
The system consists of the Brain, Spinal Cord and Nerves

Symptoms
Weakness in limbs or grasp on one or both sides
Loss of speech (Aphasia)
Inability to use appropriate words (Dysphasia)
Loss or reduction in senses
Numbness, pins & needles or altered sensation
Loss of balance (Ataxia)
Headache, Vertigo
Loss or decrease in Level of consciousness
Dizziness & blackouts
Seizures

Digestive System
The Digestive system runs from the mouth to the anus, consisting of the mouth, oesophagus, stomach, intestines, liver, pancreas, gall bladder. Its purpose is to process food to make nutrients in food available to the body.

Symptoms
Abdominal pain
indigestion
Blood in Vomit (haematemesis)
Blood in Faeces (melaena)
Weight loss, Loss of appetite
Jaundice, dark urine, pale Faeces
Difficulty in swallowing (dysphagia)
Diarrhoea or Constipation
Recent changes in bowel habits
Rectal bleeding or +/-mucus

Genitourinary System
This system consists of the kidneys, ureters, bladder and reproductive organs. The urinary system regulates the amount of fluid in the body and disposes of waste dissolved in urine. The kidney also produces some of the chemicals the body needs.

Symptoms
Loin or lower abdominal pain
Blood in Urine (haematuria)
Unusual coloured Urine
Urgency or incontinence
Painful urination (dysuria)
Impotence or loss of libido
Urethral discharge
Menstruation Problems

Labour, Miscarriage
Painful sexual intercourse (dyspareunia)
Vaginal bleeding or Discharge

Endocrine, Lymphatic and Skin
The endocrine system provides and distributes hormones needed by the body to regulate its functions. The Lymphatic system carries fluid to and from tissue spaces. Lymph nodes exist throughout the body, mainly in the torso, but also in the neck, armpit and groin. These are the glands that swell up when you have an infection. The Skin has main functions it protects the body, gives it shape, regulates temperature and absorbs substances.

Symptoms
Heat or Cold intolerance
Sweats, Shivering and Rigors
Rashes, Itching and Lumps

Musculoskeletal System
This system consists of the bones of the skeleton, muscles, joints, cartilage, tendons, ligaments. It supports the body, allows movement, provides stability and protects underlying organs

Symptoms
Pain and/or stiffness in any Bones, joints or muscles

Cardiovascular and Respiratory Assessment

The heart and lungs both reside in the chest and as there functions are so interlinked their assessments are also joined.

Questioning

Any pain in chest?
What's the pain Like? (Sharp or Crushing)
Is pain affected by breathing?
Do ankles ever swell?
Are you breathless at night?
How much exercise before breathless?
Do you ever feel faint or dizzy
Do you have a cough?
Is it productive? What's the sputum like?
Do you ever wheeze?

Examination

Before touching the patient observe their skin colour and moisture a flushed, sweaty patient may have a fever. A pale patient may be shocked. Blue tinged fingers, nose and lips (Cyanosis) shows a serious circulation problem and is often a late sign off an ongoing problem.

Listen to their breathing without a stethoscope.
A wheeze on expiration may indicate asthma or COPD. Stridor a rattling on inspiration indicates a blockage or swelling in the throat.

Record the Heart Rate, Blood Pressure, Respiratory rate, Oxygen Saturation, Temperature and Colour all are an indication of the state of perfusion which is the amount of blood reaching the tissues and organs. A pulse oximeter will measure the pulse for you but its no substitute to feeling it yourself. This will tell you the depth, strength and regularity as well as the rate.

When questioning the patient are they able to talk in full sentences or do they need to pause between words for breath. Patients who are anxious often think they are short of breath but if they are well perfused and able to hold a conversation they are usually not.

Sit the patient up slight reclined backwards, if they try to sit more upright or lean forward it can indicate they are repositioning themselves and using accessory muscles in the neck and shoulders to improve their breathing

Examine veins in neck if they are distended with blood it can indicate one of a number of conditions such as; Cardiac Tamponade, Pericarditis, Pulmonary Embolism, Right Heart Failure, Cardiogenic or Obstructive Shock and ventricular tachycardia.

Expose the chest and look for Retractions which show as skin being drawn in between ribs, above sternum and under rib cage. This shows severe difficulty in breathing.

Look for bruises and wounds which can indicate underlying damage, observe breathing as in trauma assessment, uneven breathing can indicate damage to the chest wall or lungs.

Feel the chest if it crunches like a crisp bag it indicate air trapped in the tissues form a punctured lung. Feel the ribs from front to back for tenderness and deformity.

Tap (Percuss) the chest, like all clinical skills practice this on well people before you have to do it on a potentially ill patient so you can tell what's normal and what's not. The correct method for this is to place the middle finger of one hand flat and firmly against the patient's chest and tap it with the end (not the pad) of the other middle finger, flicking the wrist to strike down.

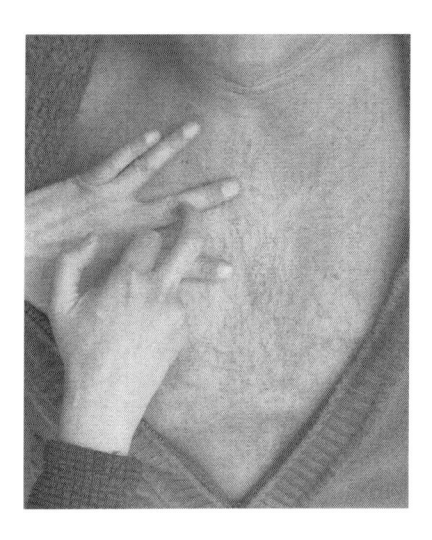

There are three tones;

Resonant (normal)
Hyper resonant (too much air, tension Pneumothorax)
Dull (Fluid or pus)

Auscultation of Lungs

Using a stethoscope listen to the chest for breath sounds (Auscultation). Start at the Front (Anterior) of

Chest and listen from side to side then move from top to bottom avoid the areas covered by bones.

Lung Sounds

Wet/ Crackles	These sounds are high pitched, they indicate the presence of infection. Mostly at the base of the lungs but sometimes throughout the lung field.
Stridor	Is a upper airway obstruction heard when breathing in.
Wheezes	Wheezes are heard in Asthma and similar diseases, in some cases they are audible without a stethoscope.
Bubbles	Bubbly sounds in the lungs are caused by fluid in the lungs in Cardiac failure
Rhonchi	These are often described as "snoring" or "gurgling" noises.
Silent	Collapse Lung or Pneumonia (Pus)
Crunching /Rub	Infection or Clot (Pulmonary Embolism)

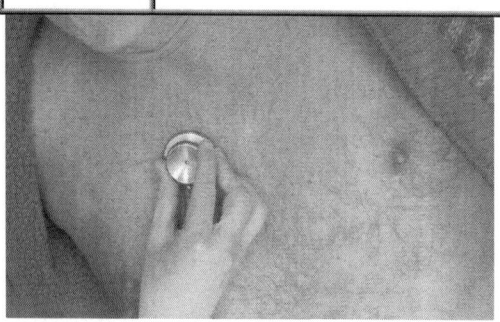

Compare one side to the other looking for differences. Note the location and quality of the sounds you hear. Can you hear air moving quietly without additional sounds in each part of the chest. Silent areas can indicate damage or infection.

Auscultation of Heart

Listen to the heart over the left side of chest. To discern all the individual heart sound and abnormalities is well beyond the scope of this text. The normal sound is referred to as "Lub dub" and is followed by a pause then repeats. Any extra noises are abnormal. A rubbing sound may indicate inflammation of the hearts sac (Pericarditis)

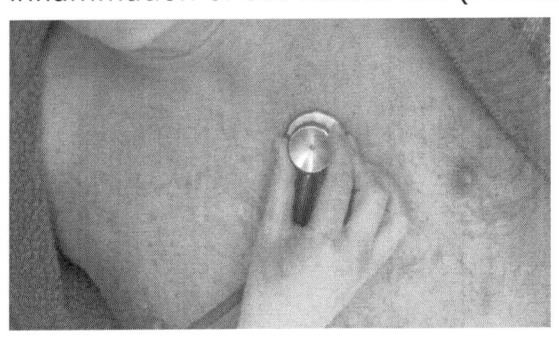

Abdominal Assessment

This covers the Genitourinary and Digestive Systems.

Questioning

Looking at the symptoms above ask about;

Presence and location of pain?

Eating Habits?

Bowel Habits?

Colour, Consistency of Vomit?

Any Fresh or old Blood (coffee grounds) in Vomit?

Colour and frequency of Urination?

If Female of Child bearing age enquire if they may be pregnant?

Examination

Ask the patient to lie flat, exposing the abdomen ask them to cough several times and observe, this may reveal the presence of a hernia. Getting them to place their chin on their chest can have the same effect.

Listen to the abdomen for bowel sounds using a stethoscope to the right of the navel. Bowels sounds are normally heard ever 5-10 seconds but listen up to a maximum of two minutes before deciding they are missing. Absent bowel sounds is an indication of peritonitis or other serious abdominal problem. If bowel sounds are frequent with a high pitched tinkling quality this can indicate an obstructed bowel, these patients would also have a history of constipation.

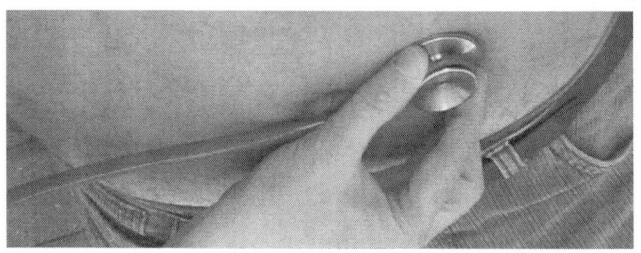

Feel (palpate) the abdomen using the fingers of a flat hand don't dig or poke the surface as this will cause muscles to contract. If the patient is conscious and tells you they have pain in a certain area leave that part to last. They may be tender in other parts which are masked by more severe pain in one location. Note areas that are rigid, tender, where you feel resistance (Guarding) or cause pain when you press and then release (rebound tenderness). Start with a gentle palpation over all regions of the abdomen to test for distension and rigidity, then press deeper to test for tenderness.

Look for scars, have they have had their appendix removed suggested by a scar in the right lower quadrant of their abdomen. Feel for a pulsating mass this can indicate an aortic aneurism.

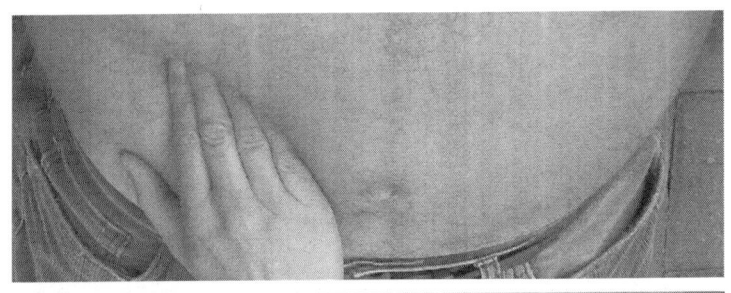

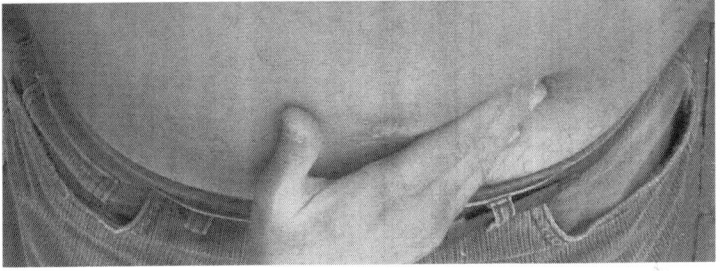

Percussion (Tapping) of the abdomen can reveal enlargement of the Liver or spleen. When tapping over an organ the sound will be duller than over the lungs or bowel. From this you should be able to identify the size of the liver and spleen. They can also be identified by palpation. The spleen is difficult to palpate unless enlarged and only a small amount of the liver is usually felt. It is also possible to feel the kidneys by supporting them from the rear and palpating from the front. All abdominal examinations require a lot of practice on well patients before a practitioner can tell what is abnormal.

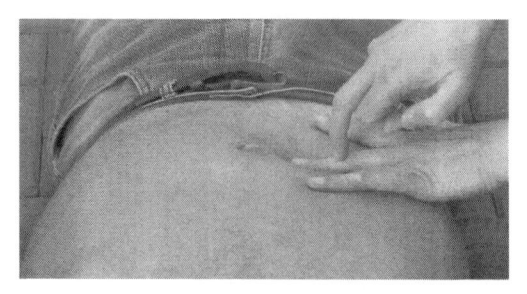

In order to reach a diagnosis and understanding of the underlying anatomy is of the abdomen is essential.

Two Pictures below used with kind permission of:
http://www.ambulancetechnicianstudy.co.uk

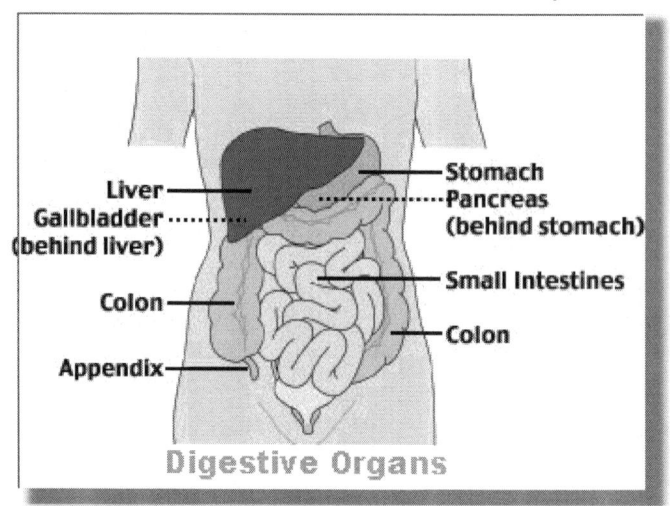

Digestive Organs

Liver
Gallbladder (behind liver)
Colon
Appendix
Stomach
Pancreas (behind stomach)
Small Intestines
Colon

When describing the abdominal regions two systems are used the first divides the abdomen into nine regions as shown below centred around the navel called the umbilical region (see below), further the sides of the body are referred to as left or right regions. So pain may be in the Right Lumbar or Left Iliac Region. The denotation is based on the patients Left and Right rather than the practitioner looking at them. These names are difficult to remember so another system is commonly used dividing the abdomen into four sections centred on the Navel. The Left and Right Upper Quadrants and the Left and Right Upper Lower Quadrants. Abbreviated to LUQ, RUQ, LLQ, RLQ. These two systems can also be mixed as seen below.

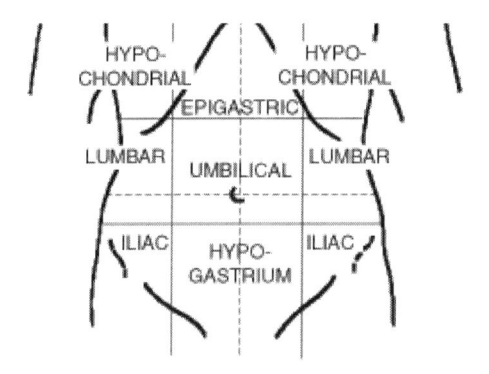

Diseases causing pain by Abdominal Region

RUQ
Cholecystitis, Duodenal ulcer, Hepatitis, Pyelonephritis Appendicitis, Pneumonia (Base of Lung)

LUQ
Ruptured spleen, Gastric ulcer, Aortic aneurism, Perforated colon, Pyelonephritis, Pneumonia (Base of Lung)

RLQ
Appendicitis, Pelvic inflammatory disease, ectopic pregnancy, Kidney stones, Strangulated hernia, diverticulitis, Crohn's disease

LLQ
Constipated, Impacted Faeces, Pelvic inflammatory disease, ectopic pregnancy, Kidney stones, Strangulated Hernia, Diverticulitis, Crohn's Disease, Ulcerative Colitis.

Epigastric
Indigestion, Pancreatitis, Myocardial infarction, Peptic Ulcer, Cholecystitis.

Umbilical
Intestinal obstruction, Pancreatitis, Early sign off Appendicitis, Aortic Aneurism, Diverticulitis.

Hypogastric
Urinary Tract Infarction (UTI)

Nervous system assessment
Questioning
To test cognitive function perform an abbreviated mental test score (amts). A score of less than 7 indicate a problem.

Give random address to remember

1. Ask Patients age?
2. Ask what time it is to nearest hour?
3. What year is it?
4. Ask who two people nearby are?
5. Ask Date of Birth?
6. Dates of second world war?
7. Name of monarch?
8. Ask where they are now?
9. Ask them to count backwards from 20 to 1

10. Ask them to recall address given at the start?

Are you unsteady on your feet?
Are you ok with heights? (Vertigo)
Any headache? Where does it hurt?
Any Dizziness or blackouts

Examination

A simple way of assessing a patient you think may have had a neurological problem is through the FAST acronym. It stands for FACE, ARMS, SPEECH, TEST.

FACE

Look for facial droop, dribbling etc

ARMS

Ask them to squeeze both your hands with theirs, feel for a marked difference in strength between body sides.

SPEECH

Listen for slurring, confusion, using inappropriate words

To expanded on the above have the patient;
- Bend their elbows and push you away then pull you towards them.

- Place your hands on their shoulders and get them to shrug.
- Bend knees and get patient to straighten leg whilst you push against them.
- With Patient on their back have them raise legs.

judging power on both sides with each test.

Sensation

To test sensation get them to close their eyes and say yes when you touch them. First use a soft object like a gauze swab then a sharp object. Check down each side of the body and check they can feel both sensations

Balance & Co-ordination

Have the patient stand the feet shoulder width apart, close their eyes then raise one leg and balance repeat on opposite side.

Holding your finger about 50cm from the patients face. Get them to move their finger from their nose to your finger, with one side then the other.

Ask the patient to run the heel of one foot, down the shin of the other leg then repeat on the other side.

Face (Testing the cranial nerves)

Note any asymmetry or drooping

Ask the patient to screw their eyes up. They should be symmetrical

Ask the patient to stick their tongue out. It should be in the midline

Ask the patient to swallow a small glass of water

Eyes (PERLA)

Look for pupil symmetry and reactivity to light

When shining light into each eye, the pupil should become smaller. Move your finger up, down, left and right have the patient follow the movement with their eyes whilst keeping their head still.

Orthopaedic Assessment

An orthopaedic assessment is usually focussed to a particular joint where the patient is experiencing pain or discomfort. To minimise the discomfort for the patient always look before you feel and feel before you move the joint.

Questioning

When did the pain start?

Any injury or event before pain?

Can you put weight on injured limb?

Has it affected your strength?

What cant you do that you could before?

Examination

Firstly compare the injured joint with the unaffected side. Look for Swelling, Deformity, Lesions and Colour changes

Feel the joint, check to see if it's hot, if the joint is swollen this may be due to an excess build up of fluid called an effusion. A small amount of fluid is normal within a joint but some diseases cause this to increase. Check to see where the joint is most tender and visualise the underlying anatomy to identify the affected structure.

Lastly mobilise the joint this is too check it has a normal range of movement. The first movement is passive allowing the patient to move the joint themselves, then actively with you moving the joint. Check the stability of the joint and supporting ligaments.

Chapter 2 Environmental Problems

Sunburn

Prevention of sunburn is better than cure, wear loose fitting cool long sleeved tops, wide brimmed hat, sun glasses and sunscreen. Too much sun can damage the skin, usually it's only superficial but severe sunburn may cause partial thickness burns with blistering. If burned;

- Protect skin from further damage
- Keep in Shade
- Encourage Cool Oral Fluids
- Apply cool moist flannel to affected areas
- Immerse affect parts in cool water
- Apply after-sun ointment

Heat Exhaustion

Heat exhaustion is caused by loss of water and essential salts from the body. The patient develops a headache and may become dizzy and confused. Show signs of shock including sweating, pale clammy skin, nausea, rapid weak pulse and fast breathing. They also may have cramps in limbs and abdomen.

- Keep in Shade
- Encourage Cool Oral Fluids
- Give Rehydration sachets
- Lie down and treat for shock
- Monitor Vital signs

Extreme Heat exhaustion can be life-threatening

Heatstroke

Heatstroke is caused by a long period of heat and can follow on from heat exhaustion. Unconsciousness can develop quickly. The patient develops a headache and may become dizzy and confused. Show signs of hot, flushed and dry skin, full bounding pulse and a fever of 40 degrees plus.

- Keep in Shade
- Remove outer clothes
- Cool with wet towels or by fanning
- Monitor Vital signs and Temperature

Dehydration

Dehydration like Heat exhaustion is caused by loss of water and essential salts from the body. A person should drink at least 2.5 litres of fluid a day, more in hot climates, to replace fluids due to heat and exercise.

The patient develops a headache and may become dizzy, confused, feel nauseous and they may also have cramps in limbs.

- Keep in Shade
- Encourage Oral Fluids
- Give Rehydration sachets
- Stretch and massage areas of cramp
- Monitor Vital signs

Hypothermia

Hypothermia starts when the body's temperature falls below 35 Degrees Celsius and can be caused by cold weather or immersion in cold water. A person will begin shivering as the body attempts to generate heat. Their skin becomes cold, pale, dry and they become lethargic. As Hypothermia progresses both breathing and heart rate slow and weaken. Once core temperature falls below 30 Degrees Celsius the patient rarely recovers and may eventually go into respiratory and cardiac arrest.

Protect Casualty from cold, wet and wind

- Prevent further heat loss by changing wet clothing
- Warm them slowly
- Cover with coats, spare blanket or survival bag
- Protect from damp ground
- Give warm drinks and high energy food
- Monitor Vital signs

Frostbite

Frostbite occurs when tissues freeze, in severe cases can lead to tissue death and the loss of digits or limbs. Most common in toes, fingers, nose and cheeks. First signs are altered sensation 'pin-and-needles' followed by numbness then stiffening of skin. Frostbite can also cause blood poisoning (Septicaemia).

- If in fingers place them somewhere warm like armpits
- Remove gloves, rings etc

- If available place in warm water 40 Degrees Celsius
- Raise part to reduce swelling

Frostnip

Frostnip is a more of a temporary discomfort but if left untreated, can eventually turn into frostbite.

High Altitude Problems

Several problems affect people at high altitude;
- Decreased Atmospheric Pressure
- Decreased Oxygen Levels
- High Winds and Cold

Oxygen Levels in Blood start reducing after 2500m, above 5500m the atmospheric pressure is half what it is at sea level. To acclimatise to high altitude may take several weeks and it's recommended that once you reach 2500m you do not ascend more than 300m a day and drop down 100m before sleeping as this will make breathing easier.

There are three conditions that affect people who do not acclimatise slowly;
- Acute Mountain Sickness (AMS)
- High Altitude Pulmonary Oedema (HAPE)
- High Altitude Cerebral Oedema (HACE)

Acute Mountain Sickness (AMS)

This can affect anybody who arrives at high altitude without acclimatisation. Effects usually take 6-12 hours to manifest. Initially AMS shows as a headache and feeling generally unwell, without an obvious cause. Other symptoms include anorexia, nausea and vomiting, fatigue, dizziness and sleep disturbances. Signs may include Peripheral Swelling (Oedema), fast heart rate, diffuse crackles whilst listening to chest and a slightly raised temperature.

The best treatment is to descend by at least 100m as this usually relieves the symptoms. Pain and nausea can be treated with appropriate medication.

If symptoms are severe also give;

- Oxygen
- Acetazolamide 250mg (Oral or IV) three times a day
- Dexamethasone 4mg (Oral or IV) four times a day

High Altitude Pulmonary Oedema (HAPE)

This occurs 2-4 days after arrival at altitudes above >2500m, fast ascents make it worse and dropping down before sleeping lessens it.

HAPE causes alveoli to swell and fill with fluid, and this reduces exchange of gases.

Symptoms include; becoming short of breath on exertion, productive cough, blood in sputum, pleuritic chest pain, breathlessness when lying down, confusion and other symptoms of AMS.

Signs are; Mild fever, fast breathing and pulse, crackles in base of lungs when listening to chest. Blueness(Cyanosis) in severe cases.

Treatment

- Descend to lower altitude
- Give oxygen 6-10L via face masks
- Give Nifedipine 20mg (By mouth four times a day)
- Give Dexamethasone as above if AMS/HACE present

High Altitude Cerebral Oedema (HACE)

HACE is rare but potentially fatal, it follows on from AMS and takes 3-4 days to manifest. It is caused by a lack of oxygen which causes the brain to swell.

Symptoms include; Headache, Nausea, Confusion, Disorientation and Hallucinations

Signs; Unsteady on feet (ataxia), Low oxygen levels (Hypoxia), unusual behaviour, Pulmonary Oedema (HAPE) and in extreme cases Coma may occur.

Treatment

- Descend to lower altitude 500-1000m or below 2500m
- Give oxygen 6-10L via face masks
- Dexamethasone 8mg (Oral or IV) immediately then 4mg ever four hours.

In some cases at high altitude over 4300m, Retinal Haemorrhages may occur, no treatment is usually required but if vision is affected descending to a lower altitude will alleviate the symptoms . High altitude cachexia (HAC) is extreme weight loss affecting people above 5000m. At these altitudes between 5000-6000 Kcal is required per day with 55% coming from carbohydrates, 35% Fats and 10% protein. Vitamin input is three times normal and high fluid intake is required.

Chapter 3 Bites and Stings

Bites

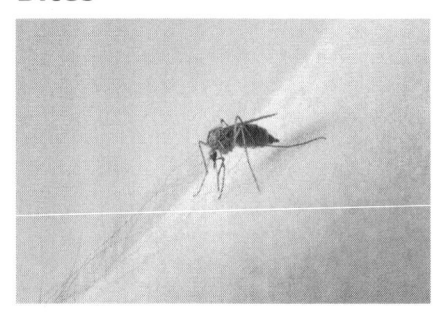

Mosquito: © Stocksnapper | Dreamstime.com

Prevention is better than cure, if in an area where biting insects are prevalent use a strong repellent that contains 30% DEET. Cover skin with suitable clothes treated with Permethrin. Natural remedies such as citronella can be used but are less effective.

A number of insects bite, including mosquitoes, midges and some other flies and bugs.

Effects vary but the following are common dependant on type;

- Local Irritation and Itching
- Raised reddened lumps
- Swelling
- Possible wound Infections
- Secondary infections transmitted via touch to eyes etc
- Transmission of disease
- Allergic or Anaphylactic reaction

Treatment depends on the severity and the likelihood of infection and complication.

Rubbing a cream or ointment such as Savlon will both soothe and help with infection prevention.

Anti-histamine creams are effective but should be used with caution as they can make the skin sensitive to light. For severe or widespread rash use oral Anti-histamines as prescribed.

For infected wounds use a topical antibacterial such as Bactoban.

If a patient shows signs such as breathing problems or loss of consciousness treat for anaphylaxis.

Stings

There are several stinging insects such as Wasps, Bees and Ants, complications are rare unless the patient develops anaphylaxis. Treat as bites above.

Ticks

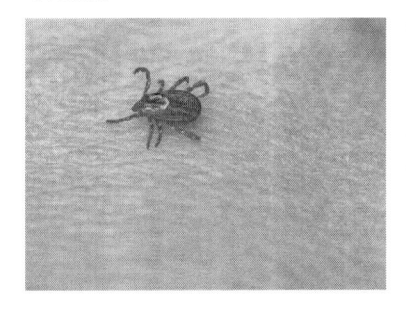

Ticks transmit diseases whilst sucking blood, but if removed within 24 hours this can usually be avoided. The best method of removing them is with tick pliers, tick twister or other specialised device.

© Melinda Fawver | Dreamstime.com

Previously recommended methods such as applying heat are now no longer recommended as this often causes the tick to inject more venom.

Poisonous Bites

In the UK there are only a few poisonous animals, but in hotter climates many poisonous species do exist.

However the risk from snake bites has often been over exaggerated. Often a bite is from a non poisonous snake, or the snake will either inject insufficient venom to harm the patient or not inject at all.

The decision as to whether to carry anti-venom on an expedition is determined by the number and prevalence of poisonous snakes in the area, and the distance and evacuation time to the nearest medical centre possessing them.

Treatment is mainly symptomatic and may include analgesia and anti nausea medication. Remove any jewellery or constricting clothing from area of wound. If poison is suspected, immobilise the patient and affected limb with splint and then bind the wound site and whole limb tightly. This prevents the lymphatic system from transporting neurotoxins to other parts of the body. Most toxins take many hours to affect the patient and binding will delay and may even prevent this. Monitor extremities for numbness and blue-tingeing (Cyanosis) to ensure bandaging isn't too tight.

Scorpion Stings

© Nico Smit | Dreamstime.com

Scorpions are found in deserts and hot, dry locations around the world.

Stings are very painful and can be fatal. Effects vary and should be treated symptomatically, with anti-venom being administered where appropriate. The pain caused by stings to fingers and toes can be relieved by a digital block and other areas by infiltration anaesthetic (see section on Local Anaesthetic)

Black Widow Spider Bites

© Dmitriy Gool | Dreamstime.com

Black Widow spider bite, starts affecting the patient within 15 minutes, starting with a dull cramping pain which then spreads to affect the whole body. Other symptoms include muscle spasms, rigid abdomen, nausea, vomiting swelling of eyelids, weakness, anxiety and pleuritic chest pain. Symptoms are debilitating but rarely fatal in a healthy adult. There is an antidote available which should be given to patients under 16 or over 65 and those with chronic medical conditions. Muscle spasm can be relieved with Methocarbamol bolus injection 100mg IV at 1ml/min, followed by an infusion of 200mg/hr iv, or an oral dose of 500mg ever 6 hours.

Lizards

Some lizards can deliver a poisonous bite, but there are no anti-venoms for these.

Fish

© Yobro10 | Dreamstime.com

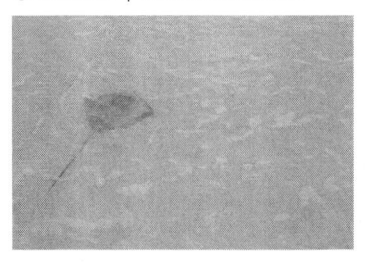

Many species of tropical fish have poisonous spines which can affect people who try to handle them, or inadvertently step on them. Commonly known species include scorpion fish, stingrays, mantas, catfish and dogfish. Stings can be very painful and deep penetration can cause serious injuries to lungs and other organs although this is very rare. Other effects, though rare may include such things as, vomiting diarrhoea, irregular heartbeat, sweating, low blood pressure, seizures and paralysis. Untreated or grossly contaminated wounds can become infected from pathogens in the water.

Stings should be removed as soon as a possible although care should be taken, as some are barbed. Immersion in warm water up to 45 degrees centigrade will ease the pain. Other methods such as a hot compress or exposure to a heat source will also help but care should be taken not to scald or burn the skin. Pain can also be relieved by a digital block and other areas by infiltration anaesthetic (see section on Local Anaesthetic)

Other marine stings

© Vaclav Janousek | Dreamstime.com

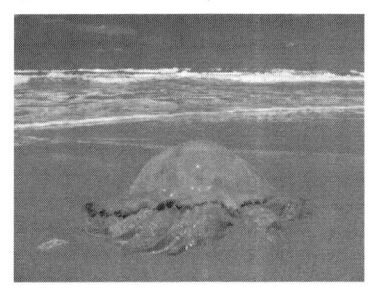

Jellyfish, Portuguese man of war, sea coral and sea anemones cause painful blisters through contact with stinging hairs on their surfaces. Effects can continue for an extended period in some people. The sting of the box jellyfish and can cause cardiac arrest within minutes - an anti-venom is available which can be given via IM injection. Other species cause severe pain, anxiety, trembling, headache, sweating, painful erections (priapism), fast heart rate, high blood pressure and fluid on the lungs (Pulmonary oedema).

Treatment to prevent further discharge of venom includes, the application of vinegar or a 50/50 baking soda/water slurry. Remove remaining creature parts with a razor, scraped across skin. Cuts from coral need to be cleaned, dressed and treated with antiseptic to prevent infection.

Chapter 4 Dealing with Parasites

Scabies

Scabies is a contagious skin infection that effects humans and other animals and is caused by a mite that burrows under the skin, causing itching and a rash.

Infection can be transferred via fabric or directly from another host and has an incubation period of 4-6 weeks. Although if cured and re-infected itching can start again within a 2-3 days. However, throughout this time, you can infect other people. Unfortunately symptoms continue for a period after mites have been killed. Infected areas often include areas between fingers, toes and in skin folds, around the genital area, buttocks and under the breasts of women. The pattern of mites burrows show as straight or s-shaped tracks in the skin, often with rows of small bites.

Treatment

Permethrin cream

Crotamiton cream

Malathion cream

In order to prevent re-infection, the patient's family and any others who have had skin to skin contact must also be treated at the same time.

Do not wash prior to applying treatment, apply to the whole body including the scalp, face, neck and ears, but avoid mucous membranes (i.e. eyes, inside the nose, or mouth).

Pay particular attention to behind the ears.

- Rub into the webs of the fingers and toes, underneath the nails.
- Treat the soles of the feet.
- Follow manufacturer's instructions for length of time of application.
- Repeat the treatment one week later.

The itching may persist for 2 - 3 weeks after treatment but will not be as intense. Eurax cream and Calamine lotion will soothe the itching.

Oral sedating antihistamine at night may also be useful to aid sleep.

Lice

Lice that effect humans come in three types;

Head Lice which are transmitted by close contact,

Patient will be itchy and both eggs and lice will be visible. Treat with repeated insecticide lotions and fine combing, promote clean hair.

Body lice spread by poor hygiene

Patient will be itchy, both eggs and lice will be visible on clothes and/or skin. Burn or sterilise clothes, bathe with soap and 1% Lysol.

Pubic Lice (Crab Lice) spread by sexual contact.

Infects all body hair, causes bluish staining and itching.

Treat Pubic Lice with repeated insecticide lotions (Permethrin, Phenothrin, and Malathion) and fine combing, and promote good hygiene

Lice eggs (nits) are oval, yellowish-white in colour, are hard to see and may be confused with dandruff. They take about a week to hatch. The empty egg cases remain after hatching.

Nymphs hatch from the nits. The baby lice look like the adults, but are smaller. They take about 7 days to mature to adults and feed on blood to survive.

Adults can live up to 30 days and feed on blood. Head lice cannot jump, hop or swim.

The itch takes from one week to 2-3 months to develop. Itching may also occur due to an allergic reaction to the bites. Sores can develop due to scratching the bites which can then become infected.

Fleas

Patients can be infected by both human and animal fleas. Rat fleas in tropical countries can transmit plague and Typhus. Their bites show in small groups and are very itchy.

Treat with steroid cream or oral antihistamines

Theadworms (pinworms)

Threadworms are white cotton like worms found on the skin around anus or in stools. They cause anal irritation and discomfort. When the person scratches the area the eggs are transferred via the fingernails to the mouth and swallowed. They can also be picked up from infected clothes and linen.

- Seeing worms, (which appear like threads) on the anus, or in the faeces
- Itching, redness and soreness around the anus
- Irritability
- Disturbed sleep

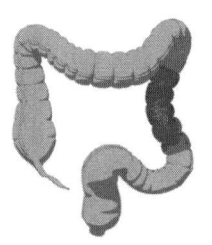

Treatment for patients over 2 years old is with Mebendazole 100 mg as a single dose; if re-infection occurs second dose may be needed after 2 weeks.

© Sebastian Kaulitzki | Dreamstime.com

Treatment for patients under 2 years old, or breastfeeding is with Senna.

Mebendazole can be used to treat Threadworm, Roundworm, Whipworm, and Hookworm infections, for adults and child over 2 years, 100 mg twice daily for 3 days.

All members of a family or group should be treated.

Worms only live for around 6 weeks and good hygiene should prevent re-infection or cross infection. Washing, vacuuming and discouraging scratching is the best way to stop spread. In very rare cases they may cause Appendicitis

Bed Bugs

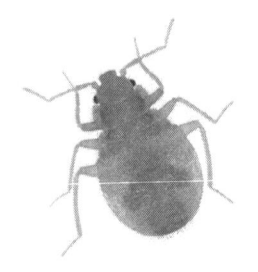

© MorganOliver | Dreamstime.com

Bed bugs do not fly, but can move very quickly. They are nocturnal, and feed at night. They can be found in crevices and cracks in wallpaper, furniture, bed frames and mattresses, even in clothes.

They are difficult to kill, bedding should be washed at the highest temperature for the fabric, and tumble dried on a high setting. Thoroughly vacuum mattresses and steam clean carpets. Powerful chemicals will be required.

Red weals appear, which are very itchy, and are often in an orderly row. Occasionally, blisters may form with swelling around the bite.

Although unpleasant, they do not carry disease. But if you are bitten and then scratch the area, this could introduce bacteria from under your nails into the wound.

Treatment

There is no specific treatment for bedbugs. Eurax cream and Calamine lotion will soothe, or an anti-histamine cream may help. If swelling occurs, apply an ice pack to the area.

Chapter 5 Foot Care

Blisters

As with all things prevention of blisters is better than cure. For feet, wearing correctly fitting, supportive, comfortable footwear and either thick socks or two pairs of thinner ones is a must. If a raw spot starts to develop, place a piece of adhesive or surgical tape over the area, this will stop friction and a blister developing.

A recent development is the use of Spenco second skin, this is a gel made from 96% water covered on both sides by cellophane. For prevention the cellophane can be left on and secured with a bandage. For treatment remove one side of cellophane and place sticky side against wound. A bandage should still be added for extra support.

If a Blister does develop and you still need to walk or use the affected part then the area needs to be thoroughly cleaned and a slit cut in to each side of blister with a sterile blade. Drain fluid from the blister and cover with either a clean dressing or by using second skin. If the latter, remove both sides of cellophane, stick one side to wound and other to bandage, keep the bandage moist, and remove the dressing after two days. Alternative dressings such as Spyroflex, moleskin or Compeed are available.

Trench Foot

When feet become wet for an extended period of time such as in swamps or jungles or they become very cold without it turning into frostbite, they can be affected by reduced circulation and become numb. The process of re-warming the feet is often accompanied by severe and long lasting pain, so the key to good management is gradual re-warming and good pain control. As the pain is neuropathic in nature amitriptyline 25-50mg is recommended.

Chapter 6 Trauma

When assessing any casualty with a traumatic injury, some important points should be kept in mind;

- Is it a fracture or Dislocation?
- Does it involve Nerves or Blood Vessels?
- Is the patient shocked?
- Does it involves any internal organs?

It is important to eliminate potentially life-threatening or complicated injuries before treating as a simple strain or sprain.

When examining the patient always **look** first, then **feel** for deformity before finally **moving** limbs or joints. If your initial Observation or tactile examination reveal a significant problem don`t move limb as this will inflect unnecessary discomfort and pain.

When performing an assessment of limbs, remove clothing and evaluate extremities using the pneumonic CSM which stands for;

(C)irculation

Is the skin a normal 'Pink' Colour or are extremities pale or cyanosed showing good circulation. Can pulses be felt away (distal) to the site of the injury. Is capillary refill normal.

(S)ensation

Can the casualty feel you touching them, is the sensation the same on both sides and are they able to distinguish between sharp and dull stimulus.

(M)ovement

Can the casualty move limbs, fingers and toes normally in a full range of motion. If pain restricts movement don't force patient to move beyond comfortable limits.

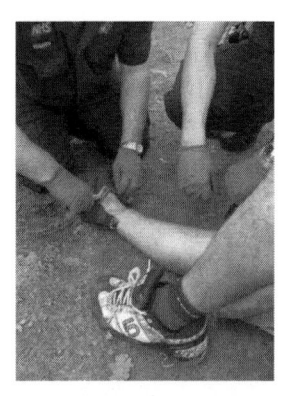

Observe for signs of shock, Be aware that multiple injuries may co-exist such as a fracture with dislocation of a joint, internal injuries and/or soft tissue damage.

Referred Pain

In the absence of a history of injury the pain the patient feels in a limb may be 'Referred Pain' this arises when the brain is confused over where the pain stimuli is coming from. E.g. nerves relating to the heart join the spinal cord at T1, T2 level this is the same area that supplies the left arm, which is why people having heart attacks sometimes complain of pain or a heavy sensation in their left arm. A pain in the shoulder with no history of injury result from a more serious underlying problem, such as pneumonia, punctured lung, aneurism, ruptured spleen or ectopic pregnancy.

Fractures

Some fractures are obvious and involve deformity of a limb or bones coming through the skin, unfortunately most are not. In the absence of x-ray facilities any injury that could be a fracture should be treated as such until proved otherwise.

Fractures are either caused by acute injury, overuse and stress or disease processes such as osteoporosis if they isn't a suitable history the injury is unlikely to be a fracture.

Fractures can be diagnosed by any or all of the following;

- Pain at the site of injury,
- Loss of movement or function,
- Swelling or bruising,
- Crepitus (grating of bones)
- Deformity including shortening, bending or twisting of a limb.

Pain and tenderness caused by fractures appear at the site of injury and around the circumference of the bone. Pain on one side only is more likely to be bruising and a fracture.

If the fracture is open part of the bone may protrude through the skin this adds the risk of bleeding and wound infection to that of the fracture. If underlying blood vessels and nerves are affected the fracture is termed complicated. One further distinction is that of stable and unstable fracture. Stable fractures occur when the bones are either not completely broken or when the ends are impacted and don`t move. Unstable fractures move independently and can cause additional injuries if not splinted securely.

When dealing with any fracture where urgent evacuation isn't possible realign the bones to the closest normal anatomical position you can achieve then splint them. If the bones are then forced to heal in this position the amount of function that returns will be dependent on the accuracy of the realignment.

Support hand, arm and shoulder injuries against the body by slings or straps as appropriate. Lower limbs should be elevated when the patient is lying down. Give analgesia appropriate in strength to the injury. Healing times are based on a patient who has good splinting, a healthy diet, rest and who doesn't have to use that limb inappropriately, not always possible in a wilderness setting.

How Bone Heal

Damaged bones can replace themselves completely given time. When a bone is fractured a blood clot forms at the site of the injury. The body then sends all the materials needed for bone replacement to the injury site and this replaces the clot. In well aligned bones a soft callus joins the bone fragments together and stabilises them. If the bone fragments are not aligned or separated union of the pieces can still take place but the join is often distorted and prone to further damage. This bridging phase takes 3 to 4 weeks. Over the next few months consolidation occurs and the soft callus is replaced by a hard bony callus although it may take up to 2 years before the bone remodels itself to its original shape.

Blood Loss with Fractures

The most serious fractures are those of 'Long Bones' damage to these bones can produce massive blood loss. Blood is lost through three mechanisms;

- Damage to large blood vessels
- Leakage from Bone Marrow
- Bleeding from muscles

The average adult has 5-6 litres of blood in their bodies, a more accurate measure is;

- 70ml/Kg Adult
- 80ml/Kg Child
- 100ml/Kg Infant

Typical long bone injuries have the following blood losses;

Humerus 500ml-750ml

Tibia 500ml-1000ml

Femur 1000ml-2000ml

Closed Pelvic Fracture 2000ml-3000ml

Open Book Pelvic Fracture 4000ml

A 40% Blood lose is classed to be Life Threatening. If you have a normal total volume of 5000ml (5L) 40% is only 2000ml (2L) which can easily be lost in Long bone fractures.

Treating Fractures

© Petar Lazovic | Dreamstime.com

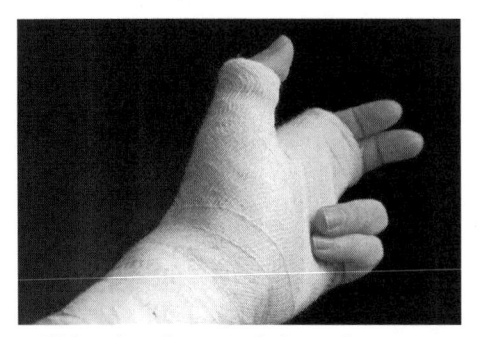

For long term immobilisation the best treatment is with plaster of paris bandage to form a cast. Do not apply a cast immediately as the limb may swell and if in a cast this will be both painful and may cause circulation difficulties.

Plaster of Paris for trauma use comes in rolls of pre-impregnated bandages and slabs. Firstly apply tubular gauze and padding to limb then evenly lay wet bandage over it to form the cast. The same techniques for bandaging a limb should be applied to creating a cast. The warmer the water using to wet the bandage the quicker the plaster dries.

A flexible splint such as a SAM Splint, can be moulded to support a variety of fractures for both short term use and where plaster is not available.

© **Caroline Jackson**

Fingers and Toes

Any fingers or toe that are fractured should be padded and strapped to the neighbouring digit. They should take 3-6 weeks to heal.

Hands

Fractures to the bones of the hand should be padded and bandages, if the injury is at or near the wrist, splinting should include the forearm to maintain stability. Fractures to the hand and or wrist can take 4-12 weeks to heal dependant on location and complexity. A flexible splint can be moulded as an improvised back slab then bandaged into place.

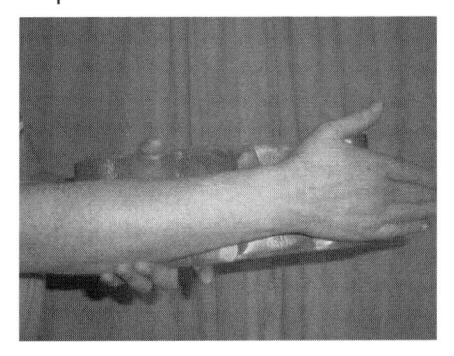

© Caroline Jackson

Feet

Fractures of the bones in the feet are difficult to confirm, adequate support can usually be achieved with a boot but the foot should be monitored in case excess swelling effects blood supply. If the damage is to the heel then it is unlikely they will be able to walk on it, Foot injuries should take between 3-12 weeks to heal dependant on location and complexity

Ankle

It is sometimes difficult to distinguish between sprains and ankle fractures. If pain persists for more than a few days and the patient is unable to weight bear treat as a fracture unless proved otherwise. If any deformity is present realign before splinting. A flexible splint bent into a 'U' shape and secured to the lower leg will provide support. If the fracture is minor the patient may be able to still walk on it once splinted. It should take around 6 weeks to heal

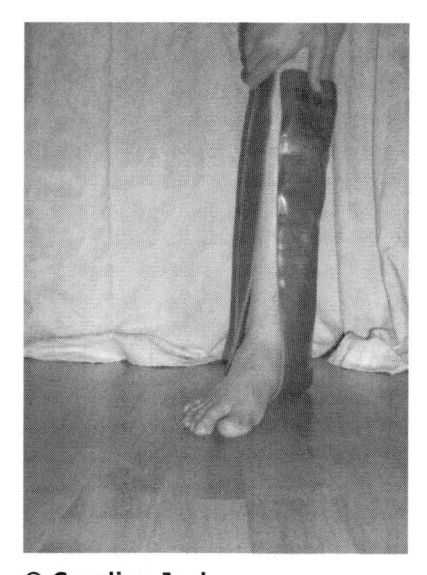

© Caroline Jackson

Forearm (Radius/Ulna)

Any fracture of the forearm is going to affect the wrist and both need to be immobilized together. Once splinted the injury can be further protected by placing in a low arm sling with a broad bandage holding it against the body. It should take around 6 weeks to heal.

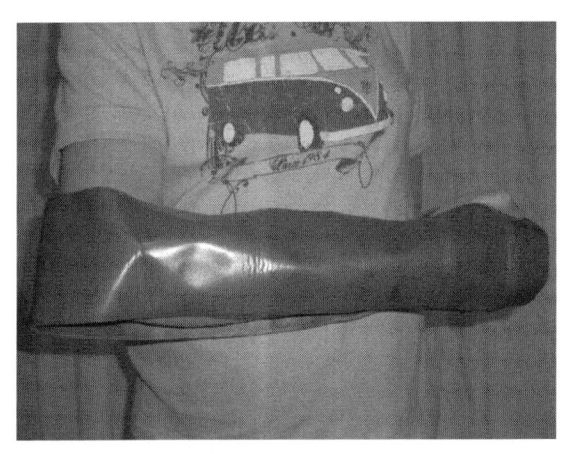

© Caroline Jackson

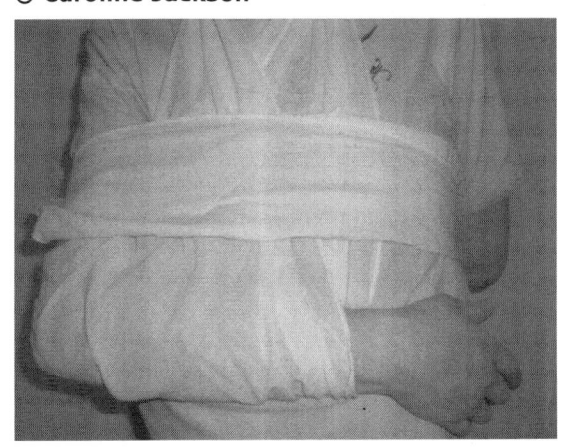

© Caroline Jackson

Elbow, Upper Arm (Humerus)

To immobilise the elbow or upper arm use flexible splint, sling and broad bandage as Forearm above.

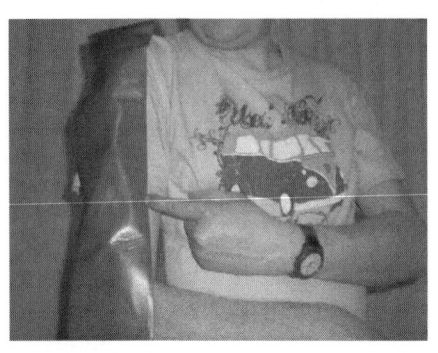

Should take around 4-10 weeks to heal

© Caroline Jackson

Shoulder Blade (Scapular)

To immobilise the shoulder use a sling and wide bandage as Forearm above

Should take around 4-6 weeks to heal

Collar Bone (Clavicle)

To immobilise Collar Bone (Clavicle) use sling and wide bandage as Forearm above

Should take around 3-8 weeks to heal

Knee and Lower Leg (Tibia, Fibula)

Lower leg fractures need to be immobilized by placing splints on either side of the lower leg. Any fractures of the knee require the ankle to be immobilized as well. Lower leg injuries will take around 10-24 weeks to heal.

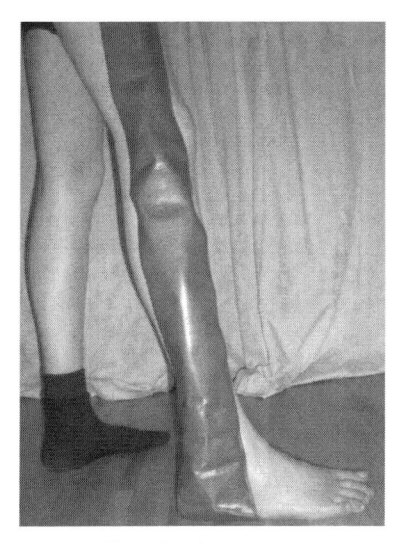

© Caroline Jackson

Kneecap

A fractures of the kneecap will need immobilization from ankle to hip with a straight splint at the back of the leg it should take around 4-6 weeks to heal

Upper Leg (Femur)

A lot of force is needed to fracture the femur, and femoral fractures are often associated with other serious complications. The strong muscles in your thigh will cause fractured bone ends to overlap and damage surrounding tissue and blood vessels leading to serious bleeding. A femoral fracture can be identified as there may be;

- Shortening of the limb
- Swelling
- Pain
- Signs of Shock

If the femoral artery is damaged death from internal bleeding can be rapid. In order to minimize bleeding traction should be applied. But this is contraindicated if there is damage to either;

- Pelvis
- Knee
- Top of Femur

There are a number of commercial traction splints available such as the Kendrick, Sager, Hare and Donway Splints. All of which cost around £100-£500.

All of the splints work in a similar way and if a dedicated splint is not available one can be improvised using Ski or walking poles, wooden staves etc

The splint provides a fixed point at the hip, additional points down a ridged frame to support the leg and an anchor point beyond the heel from which traction can be applied to a strap around the ankle. If the splint needs to be left on for any time the knee needs should be slightly flexed to avoid damage. The ankle needs to be padded to prevent damage to the circulation of the foot.

The injury is serious and will take around 12 weeks to heal.

Hip

Damage to the ball or socket of the hip joint is very painful and usually requires surgical repair. It is often diagnoses by an obvious shortening of the effected limb and an external rotation of the foot.

Pelvis

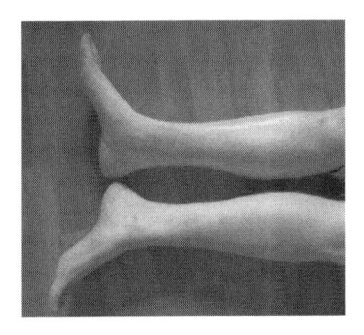

Fractured pelvis should be suspected with any high velocity injury such as falls or RTCs. If the pelvis is entirely broken the patient feet will both be externally rotated this is known as an Anterior-posterior or open book fracture.

The main complications of pelvic fracture is major blood loss as mentioned above there are major blood vessels passing through the pelvis to supply circulation to the lower extremities. Other problems can occur if splinters of bone penetrate abdominal organs. Stable fractures should take around 4-6 weeks to heal. The pelvic can be stabilized by a wide belt such as used for climbing or a purpose made pelvic splint as below.

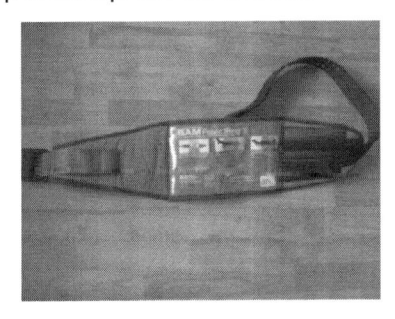

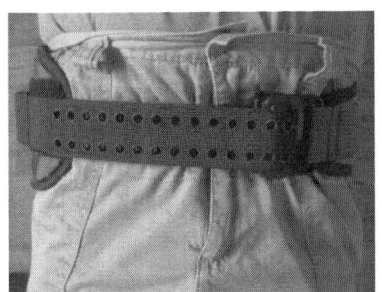

Due to the forces needed to fracture a Pelvis the patient often has other serious injuries.

- 50% Serious Head Injury
- 50% Long Bone Fracture
- 20% Serious Chest Injury

There are different types of Pelvic Fractures;

- 60-70% Lateral Fractures
- 15-20% Open Book Fractures
- 5-15% Vertical Fractures

The most common type is Lateral Fractures mostly caused by side impacts these rarely require operation and heal with bed rest. The other two share a 6% mortality rate and are often associated with severe bleeding.

Ribs

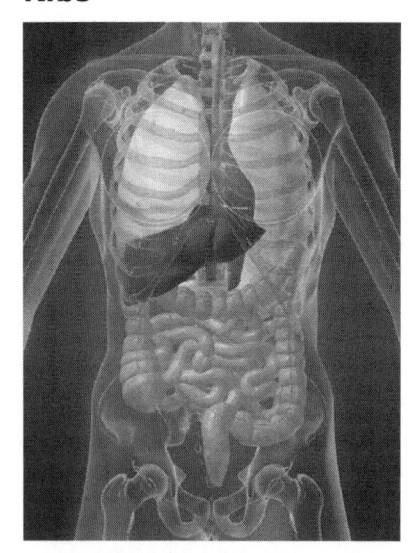

© Sebastian Kaulitzki | Dreamstime.com

Fractures to the ribs themselves usually heal without further intervention and should take around 4 weeks. The only treatment normally needed for simple rib fractures is pain killers such as a non-steroidal anti-inflammatory such as ibuprofen. Strapping of the chest is no longer recommended as it decreases ventilation and can be an increased risk of chest infections.

Fractures to ribs can interfere in ventilation of the chest. If multiple adjacent ribs are broken in more than one place a 'flail segment' is created. The segment moves independently from the rest of the chest wall causing pain and impeding respiration. Pieces of ribs can penetrate the lungs causing a heamothorax (Blood in the pleural cavity) or a pneumothorax (Air in the pleural cavity). See section on chest injuries for more details.

Spinal Damage

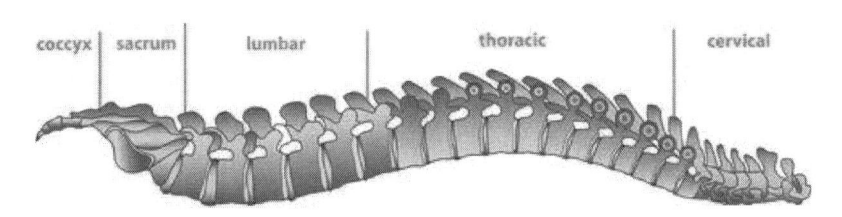

© Maryna Melnyk | Dreamstime.com

The spine is divided into five sections the cervical, thoracic, lumbar, sacrum and coccyx. Most spinal injuries occur in the cervical spine. This is mainly as it is stressed as the head is shaken due to sudden acceleration and deceleration forces which may be applied during traumatic injuries such as RTCs, blows and falls from heights.

The cervical spine is made up of seven cervical vertebrae separated by inter-vertebral disks and joined by ligaments. They are referred to as C1 to C7 starting from the base of the skull through to the top of the thoracic spine. Damage off C1 or C2 are often fatal if the spine is compromised as the nerves joining the spine at this level control breathing.

A set of rules to eliminate serious injuries to the C Spine is called 'clearing the C Spine'

If the following apply

- Fully Alert (GCS 15)
- Not Intoxicated
- No Distracting Painful Injury
- No Neuro signs or Symptoms such as 'Tingling'
- No tenderness in the Midline of the neck

And if any of the following are relevant;

- Neck Pain is present but has a delayed onset
- Has walked unaided since injury
- Neck injury followed a rear end shunt

Poor spinal management can have devastating effect on a casualties life and recovery if in any doubt about the presence of a spinal injury immobilise the casualty particularly in the following circumstances.

- Fallen >1m
- Diving accidents
- RTC>60mph
- In rollover RTCs or ejected from vehicle
- Pedestrian or Motorcyclist in RTC
- Patient over 65

There are many different types of spinal fractures but the classification is irrelevant in the wilderness setting. Any casualty that is unconscious or has a head injury should be considered to have a cervical spinal injury until proven otherwise. These casualties should be immobilized.

The first step is manual immobilisation and involves holding the patients neck in a neutral position.

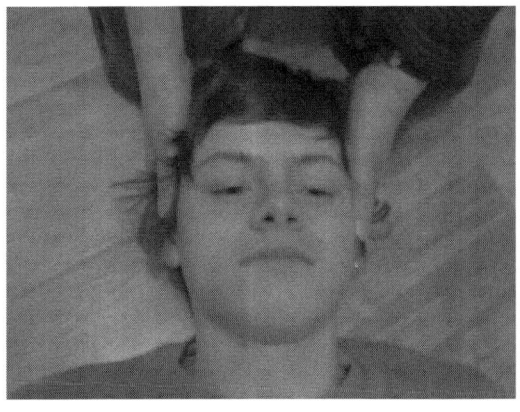

© **Caroline Jackson**

Then a collar should be applied. You can get sets of collars which are different sizes, a more versatile adjustable collar is available which fits most adults.

To correctly fit the collar measure the distance between the muscle at the top of the neck and the angle of the patients jaw. The easiest way of doing this is by placing hand and counting the number of fingers separating the two landmarks.

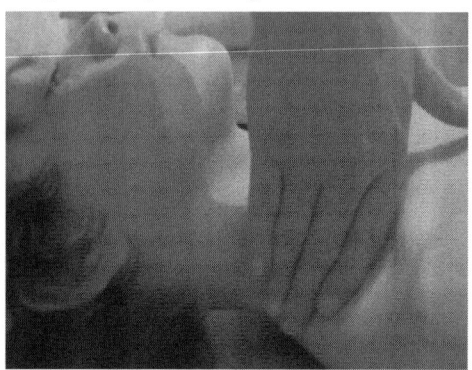

© Caroline Jackson

Compare this with the sizing line on the collar, moving the adjustable panel up until the red marker shows above your fingers. Lock the collar in the desired position using the two clips

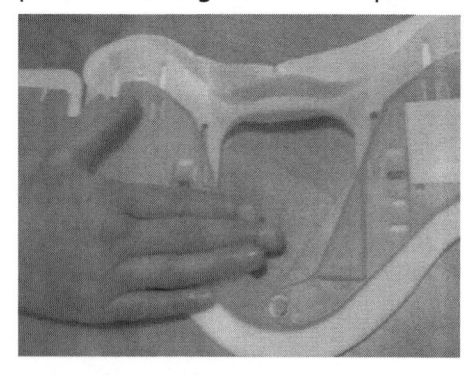

© Caroline Jackson

Slide the collar under the patients neck.

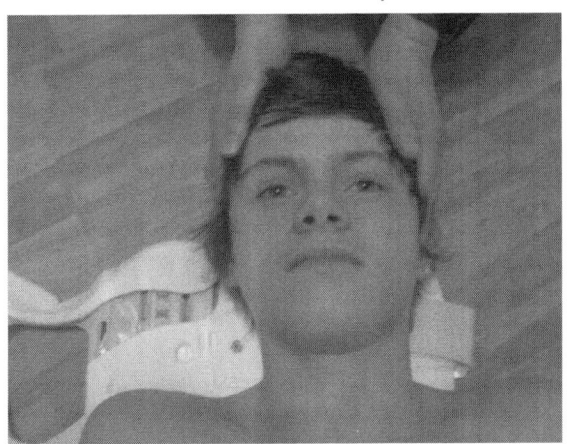

© **Caroline Jackson**

Fold over the neck of the collar over and secure it with the Velcro strap.

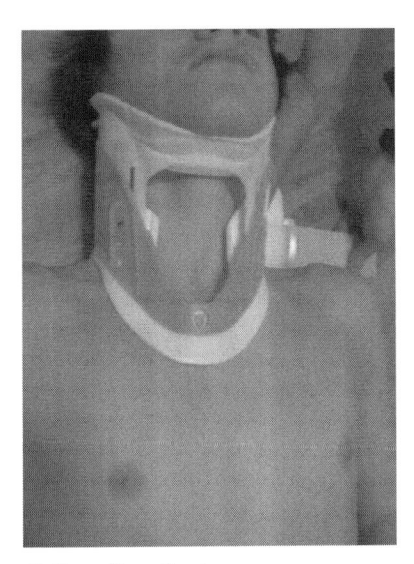

© **Caroline Jackson**

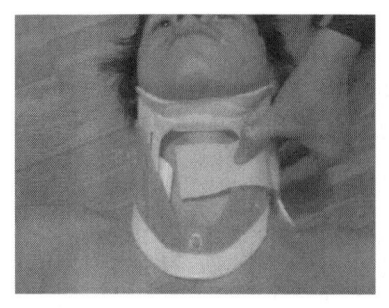

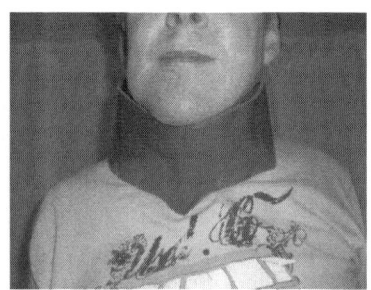

© Caroline Jackson Improvised Collar using Flexible Splint

The collar on its own doesn't provide complete immobilization the patient should then be secured on to a long (Spinal) board or for a more complete immobilization a extrication device is firstly applied.

To secure a patient to a board you need three people.

© Vatikaki | Dreamstime.com

Immobilisation of patient on Floor

1. Fit cervical collar to patient as shown
2. Lay the patient on their back with arms and legs straight

3. Place the long board next to the patient

4. Log roll the patient this should be done with 3 people one at the head who keeps it in a neutral position and two who roll the patient on to their side

5. If the patient needs to be repositioned on the board they are moved up or down in a straight line, never side to side which can place strain on the neck

6. Restrain the patient's torso using cross straps going over each shoulder, crossing the chest and being secured near the opposite hip.

7. Restrain the hips and knees with single cross straps.

8. Restrain the feet using a figure of eight strap.

9. Place Foam Blocks, Pad or rolled up towels each side of the patient's head and secure with straps or tape.

10. Check all straps are secure, breathing isn`t restricted and reassess your patient.

Immobilisation of standing patient

This is called a 'Standing Take Down'

1. One team member stands behind patient and hold head

2. Fit cervical collar to patient as shown.

3. Place long board behind the patient

4. Other two team members reach under the patient's armpits and grasps the spine board

5. Tilt the board backward and slowly lower the casualty to the ground

6. If the patient needs to be repositioned on the board they are moved up or down in a straight line, never side to side which can place strain on the neck

7. Restrain the patient's torso using cross straps going over each shoulder, crossing the chest and being secured near the opposite hip.

8. Restrain the hips and knees with single cross straps.

9. Restrain the feet using a figure of eight strap.

10. Place Foam Blocks, Pad or rolled up towels each side of the patient's head and secure with straps or tape.

11. Check all straps are secure, breathing isn`t restricted and reassess your patient.

Spinal Shock

The quality of treatment and immobilisation a casualty get immediately after an injury and for the next 8 hours has a significant effect on recovery of function. Poor management can worsen injuries considerably.

After the initial injury the spinal cord swells. Treatment to reduce inflammation of the cord helps to prevent further nerve damage

Immediate spinal surgery is not an option in a remote or survival situation. However some patients recover with immobilization alone, this is usually achieved with a fixed head brace called a 'halo' but would need to be improvised a patient will need two months complete bed rest for this method.

Spinal shock starts within a few minutes of an accident but may take several hours to attain maximum effect.

When the spine is 'in shock' the nervous system loses its ability to transmit nerve impulses, this usually lasts 4-6 weeks (sometimes several months) and some function can return once the shock has passed

The effects of the shock will depend on the level at which the injury occurs. But often effects sensation and movement below that level and related organs. However some function can return in the first few weeks after an injury. So it is impossible to predict the final outcome for any patient at the time of injury.

Once the patient is stabilised, care focuses on supportive care and rehabilitation. This care might include helping the patient bathe, dress, eat, change positions to prevent bedsores etc

Depending on the level of the injury initial disability will include the following. If the injury is at C8 for example the patient will also loose function of systems below

C4 and C5	Diaphragm (Breathing)
C5	Shoulder and Elbow Muscles
C6	Bending the Wrist

C7	Straightening the elbow.
C8	Bends the fingers
T1	Spreads the fingers
T1 –T12	Chest wall & abdominal muscles.
L2	Bends the hip
L3	Straightens the knee
L4	Pulls the foot up
L5	Wiggles the toes
S1	Pulls the foot down
S3,4 and 5	Bladder, bowel, anal, genitalia other pelvic muscles

Helmet Removal

If attending a motorcyclist with a full face helmet this should be removed if you can see or suspect a head injury, if the patient is having airway or breathing difficulties or if they are unconscious. To do this safely requires two people.

The first person stabilises the patients head by holding either side of the helmet.

The second person places the thumb and first finger of one hand on the patients cheeks under the helmet and the other hand under the patients neck to support the head.

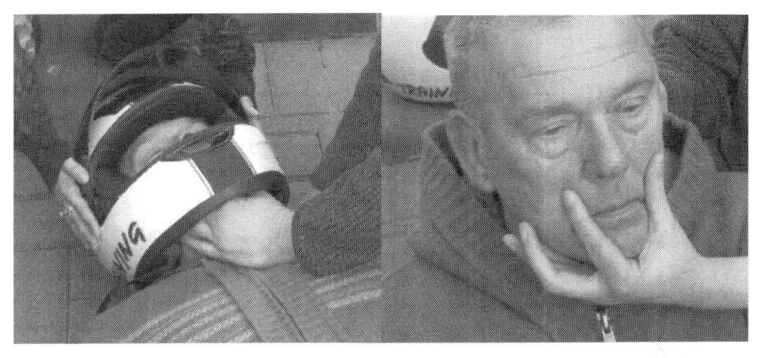

The right hand photo shows finger position without the helmet.

To remove helmet the first person, rocks it back and forth to clear the nose and the back of the skull.

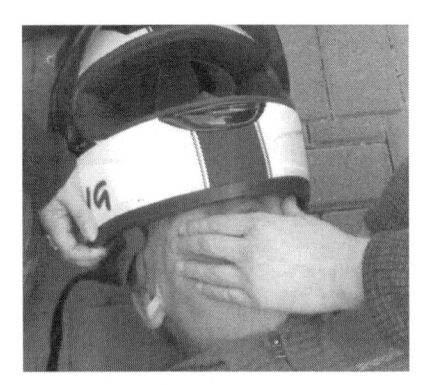

Once the helmet starts to be removed the second person can slide their hand up to support the weight of the head and reposition it in a neutral position.

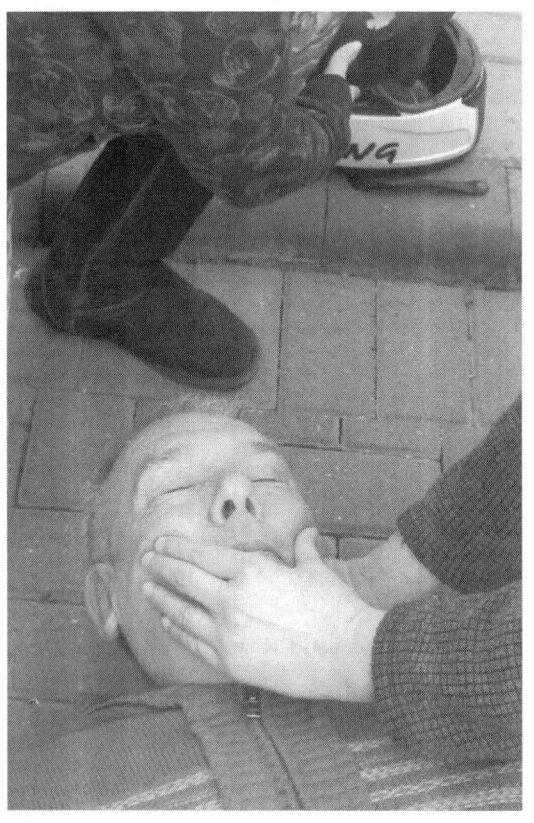

Dislocations

Before attempting to reduce any dislocation ensure that the patient has received adequate analgesic and sedatives if required.

Analgesia

Fentanyl

0.5-1 mcg/kg/dose IV/IM

Fentanyl is an excellent sedating analgesic as it only lasts 30-60 min and patient remains conscious, is reversible by narcan if required.

Morphine

Starting dose: 0.1 mg/kg IV/IM/SC

Maintenance dose: 5-20 mg/70 kg IV/IM/SC every 4 hours

Morphine is an excellent analgesic and reversible by narcan if required.

Sedation

Most patients with dislocations are very anxious , using sedation calms them and allows a lower dose of analgesic to have the same effect.

Diazepam 5-10 mg PO/IV/IM or Lorazepam 1-4mg

Jaw

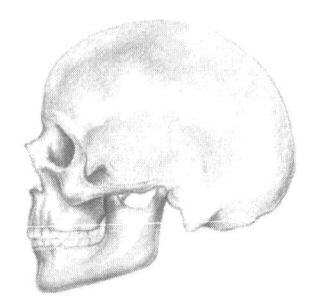

© Ciska76 | Dreamstime.com

Jaw (Mandible) dislocation is the displacement of the mandible from the rest of the skull, dislocations can be caused by traumatic and non-traumatic means. The point at which the two join is the temporomandibular joint (TMJ).

The jaw can dislocate in any direction, but moving forward anterior (forward) dislocations are the most common, Posterior (Backward) dislocations can be caused by a direct blow.

Patients have pain and are unable to open or close mouth, you should check for damage to the mouth and the stability of the jaw for fractures. Other symptoms include a misaligned bite, difficulty speaking, dribbling, jaw may be sticking out

Administer analgesia and muscle relaxation prior to treatment. Use a combination of intravenous sedatives and analgesics. Local anaesthetics can be injected directly in the TMJ space at the site of the preauricular depression. Consider midazolam as a muscle relaxation.

Sitting Reduction of Jaw

The patient is sits with you facing him, wrap thumbs with gauze this lessens likelihood of injury to your thumbs as the mouth closes after the jaw is repositioned. Put on gloves over the gauze and place thumbs across lower molars as far back as possible. The rest of the fingers are curved under the jaw.

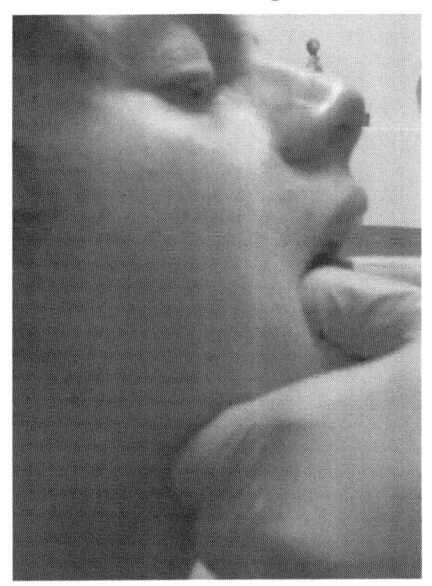

© Caroline Jackson

Classic reduction technique.

The physician places gloved thumbs on the patient's inferior molars bilaterally, as far back as possible. The fingers of the physician are curved beneath the angle and body of the mandible.

Apply downward and backward pressure on the jaw using your thumbs while slightly opening the mouth. This helps reposition the jaw.

Repositioning the jaw with patient lying on back

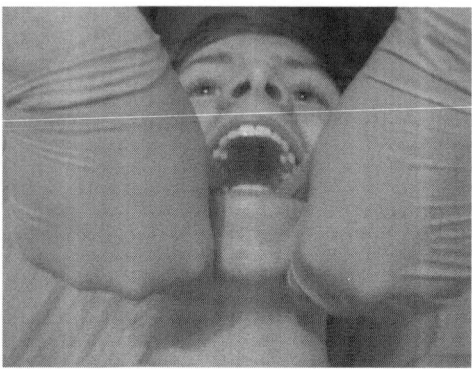

© Caroline Jackson

Recumbent approach.

The patient is placed recumbent, and the physician stands behind the head of the patient. The physician places his or her thumbs on the inferior molars and applies downward and backward pressure until the jaw pops back into place.

Places thumbs on the lower molars and applies downward and backward pressure until the jaw is repositioned correctly

Shoulder

Shoulder are the most common joint that becomes dislocated and the easiest to rectify.

When examining the patient the normal round symmetry of the shoulder looks squared off, they will hold arm in the most comfortable position and resist any attempts to move it due to pain. Check for underlying fractures and nerve and blood vessel damage. If any exist it significantly increases the risks of reducing the dislocation.

If the shoulder has been dislocated for the first time, substantial force will be required to reposition it, if the patient has had numerous dislocations the force requires lessens proportionately. There are several techniques and no single one will work in every circumstance, so its beneficial to know a few. The key to success is adequate analgesia and muscle relaxant.

If the shoulder relocates successfully the rounded contour of the shoulder will return you may also hear the head of the humerus returning to the socket. The patient will feel an instant relief from the pain and be able to move their arm as normal. Several methods exist for reducing dislocations see below;

External rotation method

Have the patient lay on their back. Firstly bend the affected arm at the elbow at a 90 degree ankle, this relaxes the tendon at the biceps. With one hand on the elbow and the other on patients wrist rotate the upper arm (Humerus) slowly which releases ligaments and allows the humerus to be repositioned. If patient experiences pain stop rotation until it settles then continues, if it fails to reduce then rotate back again it was reduce on the return. The process can be repeated or another technique tried if it doesn't work first time.

Stimson manoeuvre

Have the patient lay on their front on a raised couch, stretcher or on the edge of an outcrop if outside, they needs to be enough height for their arm on the affected side to hang down but not touch the floor. Attach a weight of some sort of between 2-5KG securely to the patient wrist this will provide continuous traction. The shoulder dislocation should reduce within 20 minutes. To help the reduction, you may apply gentle external rotation of the extended arm, flex the elbow to 90°, or manipulate the scapular.

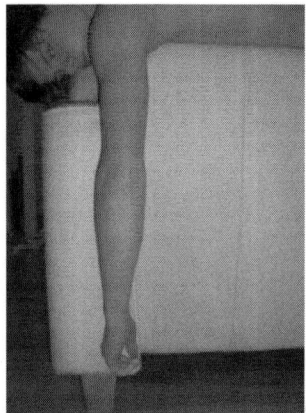

© Caroline Jackson

Scapular manipulation

The patient sits upright in a position where both back and front are accessible. Have your assistant told the patients arm out in front of them externally rotated (Palm up) and apply slight pressure whilst placing their other hand on the patients collar bone for support.

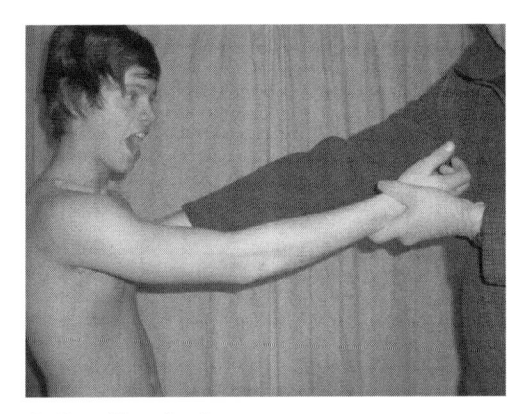

© Caroline Jackson

From the back support the shoulder blade as shown in the picture below.

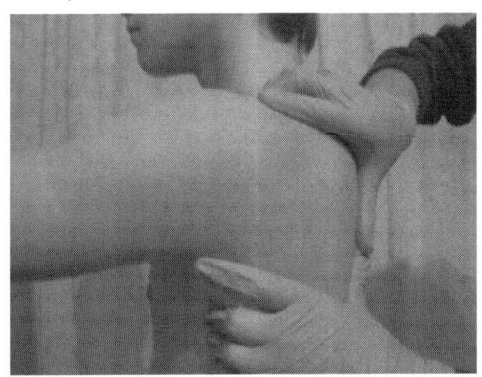

© Caroline Jackson

When the patient relaxes, use your hands to rotate the lower tip of the shoulder blade towards the spine whilst moving the upper shoulder blade towards the patients side.

To aid the reduction, the assistant may apply, along with traction elbow flexion to 90 degrees.

Elbow

The elbow is a sturdy joint and takes a considerable force to dislocate. Due to this a significant amount of elbow dislocations also have associated fractures. The elbow can be displaces to the front (Anterior) or more often the back(Posterior).

Manipulation for Posterior Dislocation of elbow

Bend the elbow to 90 degrees with hand facing backwards, push the humerus backwards whilst your assistant exerts downward pressure on the forearm. A clunk is felt and heard as the joint realigns.

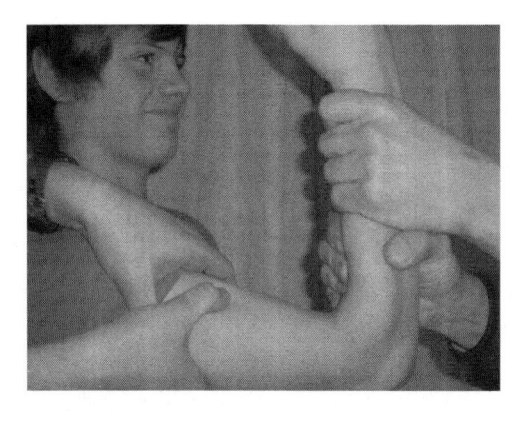

Manipulation for Anterior Dislocation of elbow

Apply gentle traction by pulling the wrist forward and downward pressure on the forearm. Take care to avoid over extending the elbow, which may cause damage to blood vessels or nerves

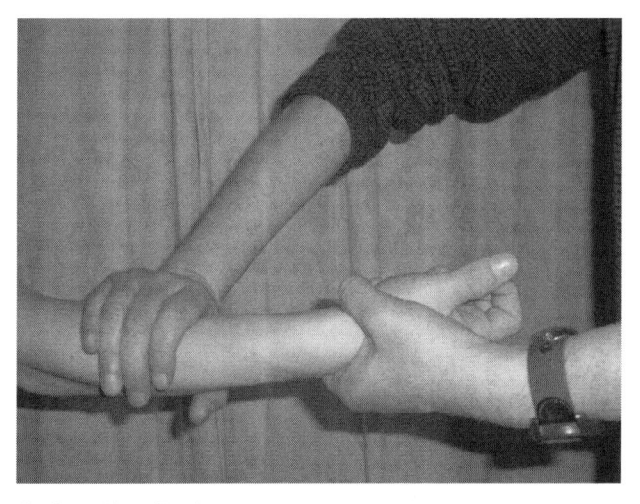

© Caroline Jackson

Finger

The two joints in each fingers are called interphalangeal (IP) joint. Where the finger joins the hand it's called the metacarpophalangeal (MCP) joint. The most common dislocation is of the proximal interphalangeal (PIP) which is the nearer joint to the body of the hand. It can dislocate to the front, back or side.

Manipulation

Apply traction and manipulate the joint to the correct anatomical position. If the dislocation is to the back splint at a 30 degree angle or if the dislocation is to the front or side splint with finger extended.

s © Caroline Jackson
Dreamstime.com

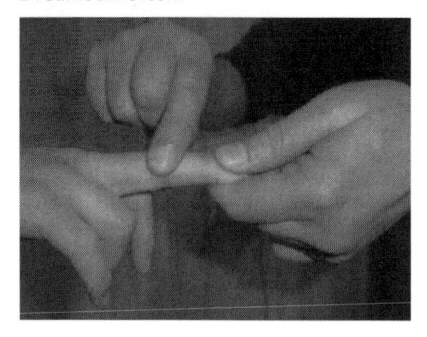

© Robert Byron |

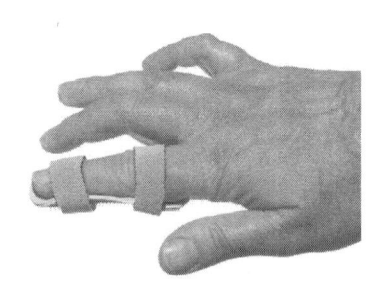

Hip

Hip dislocations can be to the back (Posterior) commonly caused by impacts in road accidents, to the front (Anterior) caused by falls or to the side (Lateral) usually associated with a fracture.

In all dislocations the patient is in severe pain, has severely limited movement in the limb. There is also a possibility of nerve or blood vessel damage. Damage to the femoral artery can cause fatal internal bleeding.

© Caroline Jackson

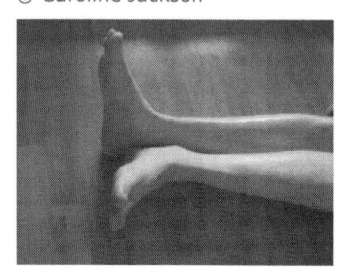

If it is an posterior dislocation the leg will be internally rotated towards the centre of the body whereas in a anterior dislocation the leg is externally rotated.

Manipulation of the Hip

Manipulating a dislocated Hip is extremely painful and adequate analgesia and sedation should be administered. However successful manipulation will relieve the patient severe pain and lessen the likelihood of complications due to nerve or blood vessel damage.

Two people are required, the patient is laid on their back. Your assistant places downward pressure on the patients pelvis to provide counter-traction. You now need to straddle the injured leg and with femur at 90 degrees and knee flexed. Bend your own knees and securely grasp the patients knee. Straighten your legs to apply traction to the limb. Further manipulation can be achieved by moving the limb from side to side.

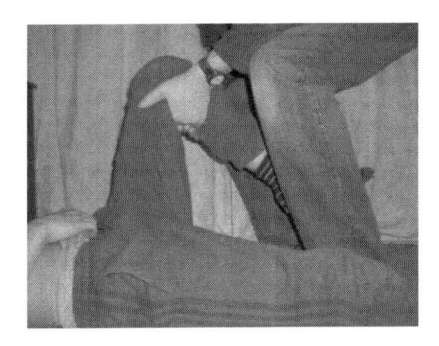

© Caroline Jackson

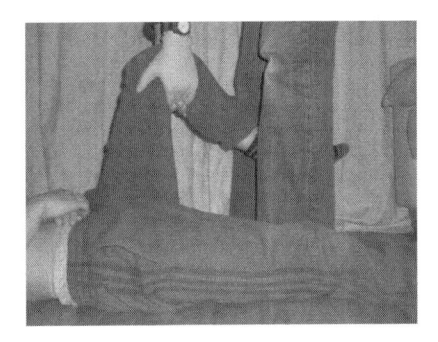

© Caroline Jackson

Once the hip is reduced splint to other leg and if possible provide light traction, the patient will then need to be transported on their back.

Knee

Knee dislocations are usually accompanied by extensive soft tissue damage, this will make reducing the dislocation easier as there will be less muscle mass working against you. Reduction is achieved by applying straight leg traction until normal position is achieved. Splint the leg in a straight position to reduce movement of the knee.

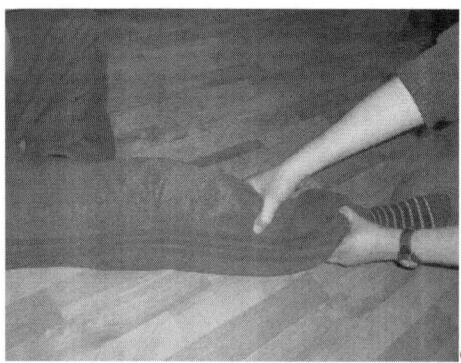

© Caroline Jackson

Kneecap (Patella)

When dislocated the kneecap usually moves to the outside and the deformity is obvious when compared with the other leg. The patient will keep their knee flexed in the most comfortable position. To reduce the dislocation straighten the leg this may be sufficient on its own but if not the kneecap can be manipulated back to the centre. If the patient is provided with sufficient analgesia he may relax enough for the kneecap to reduce on its own.

© Caroline Jackson

Strains and Sprains

Sprain

A sprain is an injury to a ligament which are strong tissues which support joints and attach bones together. They can be injured, by being stretched during a sudden pull.

Grade I - mild stretching of the ligament without joint instability.

Grade II - partial tear of the ligament without causing joint instability.

Grade III - complete tear of the ligament with instability of the joint.

A damaged ligament causes swelling, inflammation, pain and bruising around the affected joint.

Strain

A strain is stretching or tearing of muscle fibres. Caused through muscle stretching or sudden contraction.

First degree strain - a mild strain affecting a few muscle fibres which are stretched or torn. The injured muscle is tender and painful, but has normal strength.

Second degree strain - a moderate strain with a greater number of injured fibres. There is more severe muscle pain, tenderness, mild swelling, some loss of strength, and a bruise may develop.

Third degree strain - this strain tears the muscle completely with a complete loss of function.

For a long time in First aid Strains and sprains were treated using treatment based on the RICE acronym. Later we moved on to a two phase treatment RICE followed by MICE where after the initial period the injury was mobilised instead of Rested. Now we have PRICE and HARM.

Protection	**No Heat**
Rest	**Alcohol**
Ice	**Running**
Compression	**Massage**
Elevation	

PRICE
Protect
Protect the injury from further damage.

Rest
Rest the affected joint or muscle for 48–72 hours following injury. Keep weight off lower limb injuries

Ice
Ice works on muscle strains and sprains, particularly in the first 48-72 hours following the injury. Ice

reduces swelling and provide some pain relief by numbing surrounding tissue. Cooling encourages warm blood to the injury site, brining oxygen and nutrients to aid in the healing process. Never apply ice directly to the skin and apply for between 10 and 30 minutes for a time. Cooling can be achieved with a chemical ice pack or a bag of rice wrapped in a towel

Compression

Compress using tubular gauze or plain gauze bandage for first 48-72hrs, ensure that the bandage supports the injury without affecting blood flow. After this time it's best to mobilise the joint but gauze can still be applied to weight bearing joints for additional support.

Elevation

Elevation is intended to reduce swelling, for lower limbs support on a stool at hip level or in bed on a pillow. For arms keep in a elevation sling. Once the swelling has stopped you no longer need to elevate the limb.

HARM

For the first 72 hours avoid the following;

Heat

Avoid heat initially as it discourages blood flow, after the first 72 hours the inflammation should have subsided and heat therapy may be soothing.

Alcohol

Alcohol causes blood vessels to get bigger (vasodilatation) which increase sub dermal bleeding (bruising).

Running

Running or any other form of impact exercise will cause more damage.

Massage

Massage will also increase bleeding and swelling. after first 72 hours, gentle massage may be soothing.

Gunshot Wounds

Introduction

Gunshot wounds can be split into two groups depending on the muzzle velocity of the weapon. Generally handguns are considered as low velocity weapons with a muzzle velocity of < 1000 ft/sec and can be fired with one hand, whereas a rifle is high velocity weapon with a velocity of greater than 1000 ft/sec.

Exceptions to this are high calibre handguns such as .357 and .44 Magnum, which are usually fired with two hands and are treated as high-velocity. And .22 Rifles and Shotguns at long range that are considered low-velocity weapons.

The difference is mainly academic as the extent of wounds caused by firearms is determined by many factors including the deformation and fragmentation of the projectile its entrance profile and the path travelled through the body.

Most handgun wounds can be treated without surgery and with careful wound care management; although the presence of fractures are likely if the projectile hit bone. Even if the wound appears clean contamination is common and prophylactic antibiotic therapy is recommended. If open fractures are present intravenous antibiotic therapy for 48 to 72 hours is preferred or a full course of oral antibiotics.

Rifles and shotguns at close range cause high velocity wounds. Due to the characteristics of high velocity rounds the chance of open fractures and wound contamination is much higher than with handguns. In high velocity injury tissues are pushed away around the projectile path this is known as temporary a cavity. The effect produces blunt trauma that extends beyond the tissue actually contacted by the projectile if the tissue retains its contractility this cavity will disappear when the tissue returns to its normal position. The pulsation of the cavity results in a strong negative pressure that draws contamination from both the entry and exit wounds along the entire wound tract.

However even with high velocity wounds if there is only a small amount of tissue damage, no fragmentation or fractures then simple wound care may be enough.

Ballistics

In addition to the range and calibre of the weapon we must also consider the type of ammunition that is used. A normal military bullet is coated with a solid metal jacket and is less likely to fragment when it passes through the body, additional fragments may occur when a jacketed bullet fragments when hitting bone or when a non-jacketed, soft-point, hollow-point, or composite bullet fragments when passing through soft tissue thus creating larger permanent cavities by mushrooming out. Fragmentation may significantly increase injury. Should the cavity at the point become plugged with clothing, wood or other debris it may fail to expand.

Some Military rounds are designed to tumble in flight and thus create a larger tract and do more damage. . Civilian rounds such as the Black Talon, Hydra-Shok, or Golden Saber are designed to ensure complete transference of the kinetic energy to the target. Whilst other rounds may achieve over penetration and pass through the body without fragmenting, thus not impart all of its energy.

A shotgun blast creates multiple projectiles creating numerous holes and significant tissue damage, especially at close range. A variant bullet is the composite round. These have a copper hollow point outer shell containing shot stabilised with epoxy resin which fragment upon hitting a surface, creating multiple projectiles once they enter the body. This dramatically increases the size of the permanent cavity.

Even blank ammunition containing powder but no bullet may cause injury or death at close range due to the gases, heat and packing released when fired.

Evaluating the Casualty

Both knowledge and practice is needed to properly assess a firearms injury, when evaluating a wound interview any witnesses to determine both the range and angle the bullet entered from and the type of weapon used. Ask how many shots were heard, as the bullets will need to be accounted for. Any bullets or fragments removed or found should be retained to try and ascertain if all parts of the round were recovered and nothing is left within the wound.

If there are any spent cartridges cases or shell casings around these can be examined to determine both the weapon used and its calibre

Carry out a full secondary survey and instigate any life saving measures as you progress. Whilst examining the body check for additional injuries as well as entry and exit wounds. Assume that there may be both fractures and neurological damage until proven otherwise.

Types of Wounds from firearms.

The primarily mechanism of injury is the path the projectile takes when it passes through the body. A shotgun loaded with shot as opposed to a solid load fired at close will produce a large entrance wound, but will usually lack the energy to penetrate the body completely. The entry wound from a rifle or handgun may be more difficult to locate as the small diameter of a projectile and elastic property of the skin will close the wound one the bullet has passed through.

A shot fired at point blank range may exhibit charring to the clothing or skin from the muzzle flash and hot gasses escaping from the barrel when fired. A stellate, or star-shaped, wound is formed at the skin with a jagged appearance to the entrance wound and sometimes, an imprint of the barrel is left in the skin.

Complications can occur from this as the gasses are transmitted into the wound making the temporary cavity larger and carrying soot and char clothing into the cavity to further contaminate the wound.

Conversely exit wounds tend to be larger and will contain whatever debris the bullet has sucked through the body. Also even if only one shot is fired the bullet may have fragmented within the body and created more than one exit wound or propelled bony fragments through the skin.

The second injuring mechanism is the temporary cavity created by the kinetic energy of the projectile This causes tissue to be force from the projectiles path, causing stretching, tearing and concussive forces to the surrounding tissue. The cavity may measure 15 times the projectile diameter. During the first five minutes, the wall collapses and reforms (pulsates) several times. This additional sheering force can damage tissue some distance from the projectile tract itself.

Wounds to the chest or abdomen will frequently incur Organ damage, observe for signs of air (*pneumothorax*) or blood (*hemothorax*) in the space between the chest wall and the lungs, which may cause one lung to collapse. Or compression of the heart (Cardiac Tamponade) by the presence of blood or fluid in the sack surrounding the heart (*Pericardium*). Or damage to the organs in the abdomen. Not forgetting to check under the armpit and in the space between the genitals and the anus (*perineum*) for entrance and exit wounds.

Treatment

With all gunshot wounds the area must be thoroughly explored to remove fragments and any remaining projectiles. The wound needs to be irrigated and cleaned, if dead tissue (4 Cs colour, consistency, contractility, and capacity to bleed) is present in or adjacent to the wound this needs to be removed. Dead tissue is removed (debridement) until surrounding healthy tissue is exposed in order to promote healing and knitting of the wound edges. If there is damage to blood vessels, exploration and repair should be performed after fractures are stabilised.

In circumstances in which contaminated dead (*necrotic*) tissue cannot be excised promptly or adequately, a prophylactic antibiotic cream, beads, solution or spray should be applied directly to the wound to prevent otherwise lethal infection.

When assessing possible nerve damage, lack of feeling or control in extremities is not always an indication for extensive exploration for as the loss may be temporary and feeling return without further intervention.

If substantial open fractures occur it will not be possible to stabilise them is the field as surgical plates, nailing or external fixation will be required. These types of injuries require the services of an orthopaedic surgeon.

The injuries consider above are mainly associated with wounds to the limbs or flesh wounds to the trunk. Deep penetrating wounds to the chest or abdomen will effect single or multiple organs, all these injuries are potentially life threatening as the loss of function of the organ, internal bleeding and high risk of infection have far reaching effects.

It's possible to survive some organ damage without restorative surgery such as a collapsed lung or damage to one kidney providing good care is given, bleeding is controlled and infection prevented. However most organ injuries would be fatal without professional care.

In all cases the following steps need to be taken (see Relevant Chapters)

- Bleeding must be controlled.
- Fractures Stabilised
- Dead tissue removed from wounds (*debridment*)
- The wound or wounds cleaned and washed out (*irrigated*).
- If necessary the wound/s should be packed or closed as appropriate.

Types of Bleeding

There are several type of blood vessels in the body, the three well known ones are Arteries, Veins and Capillaries.

Capillaries

These are the smallest vessels and supply a vast network of blood to the skin and organs. If a person is struck and a bruise appears this is actually the result of Capillary damage releasing blood under the skin, if you graze your skin then the exposes blood is due to damages Capillaries and tends to ooze out. Capillary bleeding is very unlikely to pose a significant bleeding risk however it does open the body up to infection so all grazes etc should be covered.

Veins

Veins are used to take de-oxygenated blood away from the tissues. The blood in them is said to be a darker red as there is less oxygen in it than arterial blood. If a vein is damaged the blood runs out steadily. Venous bleeding is life threatening if not treated particularly if internal or in an unconscious casualty that cannot care for themselves.

Arteries

Arteries carry oxygenated blood from the heart to the tissues. Blood from arteries is forced out with each beat of the heart, hence the faster the heart rate the quicker the patient losses blood. Arterial bleeding is the most life-threatening but can be managed using proper techniques.

Dealing with Wounds

A useful acronym for dealing with wounds is PEEP.

(P)osition

Reassure the casualty and get them to sit or lay down depending on the site of injury. This will calm them down, lower their heart rate and thereby lower the amount of blood that may be pumping out of there wound each minute, particularly if the bleeding is arterial.

(E)xpose and Examine

Check the wound for severity of bleeding, contamination and any imbedded or penetrating objects. If there is loose glass, gravel etc on the surface of the wound brush it away. If applying immediate first aid with the intention of contacting health care services soon leave any imbedded objects in place and apply padding to protect them from pressure of subsequent bandaging.

(E)levate

If able get the casualty to elevate their own injured limb, this will free you to gather first aid material.

(P)ressure

Apply dressing and direct pressure to wound unless imbedded objects are still in place in which case apply pressure to each side of the object. The use of direct pressure is often underestimated, as even arterial bleeding can be staunched with the application of pressure.

Application of Dressings

Most people would agree its fairly straightforward applying a a dressing to a wound. There are two main ways of covering a wound. You can either cover it with a pad then apply a bandage or tape on top of the pad to hold it in place or use a readymade dressing consisting of a pad with attached bandage.
Readymade dressing come in a variety of sizes depending on type and manufacturer. Standard sizes are;

X-Large Dressing	27.5x20cm
Large Dressing	18x18cm
Medium Dressing	12x12cm
No 1 Ambulance Dressing	12x10cm
No 2 Ambulance Dressing	20x15cm
No 3 Ambulance Dressing	28x20cm
No 4 Ambulance Dressing	32x20cm
Military Field Dressing	10x19cm
Military Field Dressing	20x19cm
Military Field Dressing	30x30cm
Emergency Care NATO	15x17.5cm

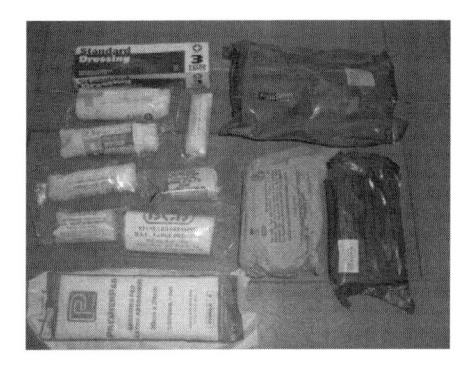

© Chris Breen

A selection dressings; Civilian on Left

Military on Right

Good principals of practice which reduces risk of infection and aid in the control of bleeding are;

Choose a dressing large enough to cover whole wound

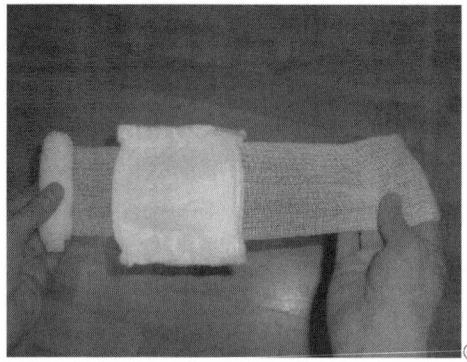

© Caroline Jackson

- Have casualty hold short end of bandage taut, allowing you to wrap dressing

© Caroline Jackson

- Apply bandage to overlap all edges of dressing.
- Tie bandage off on top of wound.

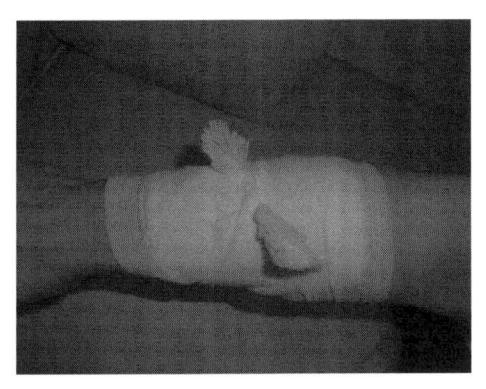

All Photos on this page © Caroline Jackson

Bandaging
Cut to Palm off hand

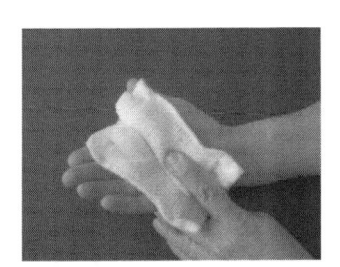

Place sterile dressing over wound.

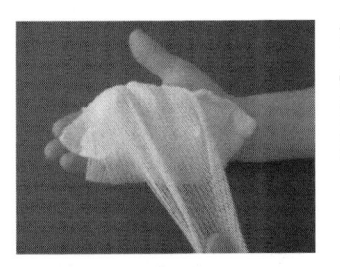

Wrap bandage or second dressing to secure it in place. Start with several turns around hand

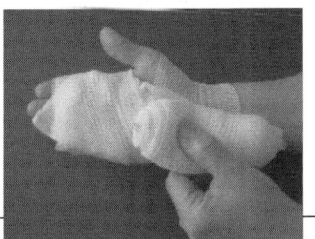

Move down to wrist and wrap bandage around limb.

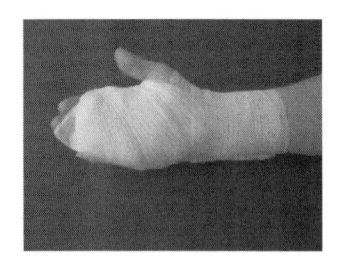

Secure end with tape. Keep thumb and finger ends visible. If they become pale, blue, cramped or numb loosen dressing.

© Caroline Jackson

Slings

There are two variety of slings commonly in usage;

Broad Arm Sling

This sling is used for support of any arm, hand or shoulder injury. To apply the sling drap a triangular bandage over the shoulder with the small point under the injured limb and the long end running down the body.

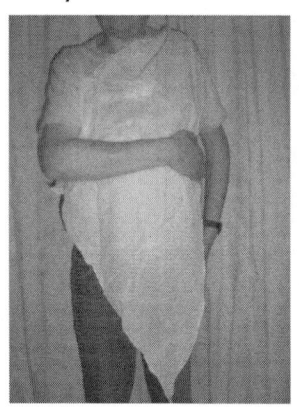

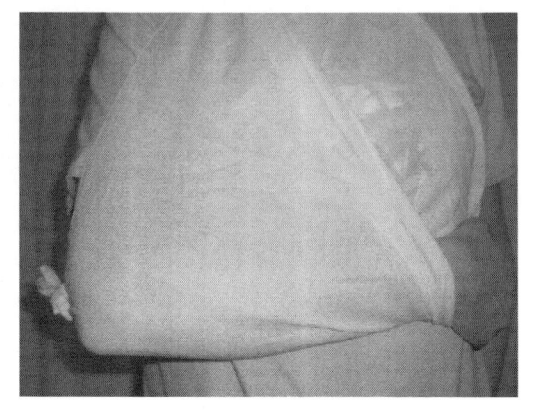

© **Caroline Jackson**

Next pick up the hanging end and place over the shoulder on the injured side and tie to other end from behind. Twist point at elbow and tie single knot to form a pocket which cradles the elbow.

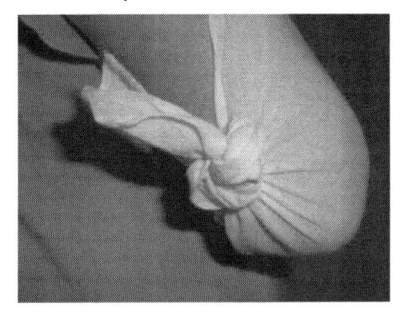

© Caroline Jackson

Elevation Sling

This is used primarily to control bleeding in hand and low arm injuries. Have casualty place injured arm across chest pointing upward towards opposite shoulder. Drap triangular bandage over arm with point towards elbow on the injured,

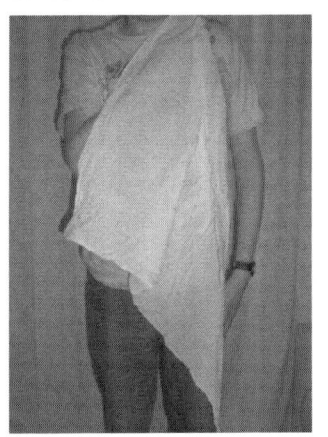

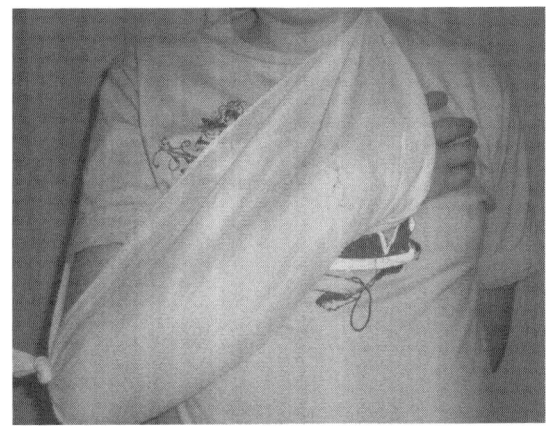

© Caroline Jackson

Bring hanging end of bandage under armpit on injured side and tie at back. Twist and tie point at elbow as above.

For both slings ensure fingers are visible to allow circulation checks to be made. If fingers start going pale or blue then dressing/bandages are too tight and need loosening. For added stabilization a broad bandage made from a second folded triangular bandage can be tied across the chest as below.

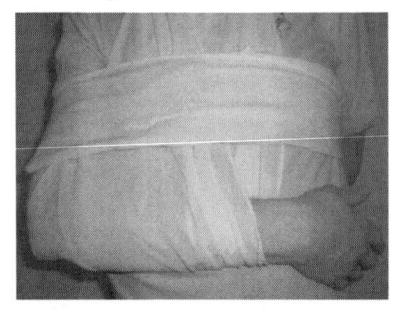

© Caroline Jackson

Control of Serious Bleeding

Most bleeding can be controlled by direct pressure and/or elevation. Where it can`t there are a number of options available.

- Indirect Pressure
- Tourniquet
- Packing wounds
- Haemostatic Agents

Indirect Pressure

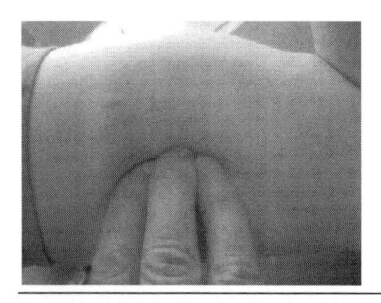

© Caroline Jackson

The two commonly used pressure points are in the arm, see photo left (brachial) and in the groin (femoral).

The brachial point is located above the elbow on the inside of the arm in the groove between the biceps and the bone. Applying pressure at this point will slow bleeding giving wounds a chance to clot. The femoral artery is located on the front, centre part of the crease in the groin. It takes a lot of pressure to stop femoral bleeding use your fist or heel to apply pressure

Packing wounds

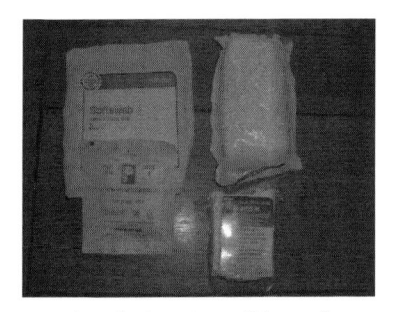

If a wound is deep it should be packed, avoid using anything fluffy like cotton wool. Instead use plain gauze squares or ribbon gauze which is useful for small deep wound as it can be fed in to the wound and packed down with a forceps.

Tourniquet

The use of Tourniquets used to be discouraged but they have proved to be very effective in recent conflicts where severe bleeding from gunshot and explosions is prevalent.

A limb tourniquet can be improvised from any band of material 5-10cms wide. It should be applied 5-10cms above the wound and above the elbow or knee if the wound is on the lower part of the limb. Leave on for 30 minutes this will give you time to identify the point of injury and apply direct pressure. Ease off the tourniquet and if bleeding continues reapply. The application of a tourniquet rarely stops circulation completely but will slow it down enough for use of other methods. The military have developed a Combat application tourniquet for just this purpose.

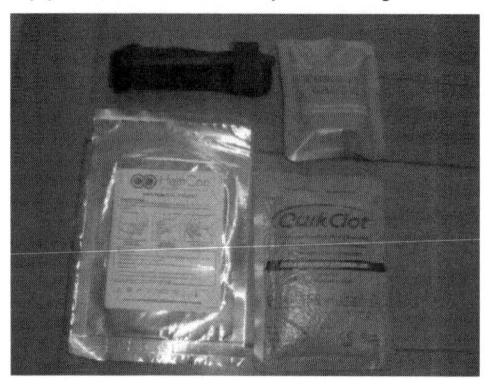

Haemostatic Agents

There are several different haemostatic agents such as Quikclot and Celox;

Quikclot was originally designed for the military although it has many civilian applications. It comes in its original form granules, Advanced Combat Sponges(ACS) and Combat Gauze. It works by absorbing the water content of blood and encouraging the bodies clotting system. The reaction cause heat to be generated which originally caused some problems this has now been changed. QuikClot must be applied to the blood vessel itself. Also contamination, dirt or accumulated blood have to be removed for it to be effective.

Celox is a similar product, made entirely from natural materials making it more easily absorbable in the body. It comes as granules, as a gauze and in an applicator for deep penetrating wounds.

Hemcon dressings

Made from chitosan the Hemcon dressing sticks to a wound and promotes clotting whilst providing an anti-microbial barrier. Comes in a variety of sizes. Expensive compared to other similar products.

Principles of Wound Management

Most wounds in otherwise healthy individuals heal by themselves or with little intervention other than initial cleaning and having a dressing applied to protect them from further damage and infection.

When a wound is deep or complicated by other factors a regime needs to be implemented to

Manage the healing process. The management plan needs to consist of the following

- To assess, plan, implement and evaluate care with consideration for the whole person not just the wound.

- To promote the natural healing process by maintaining a warm, moist and non-toxic environment.

- To use treatments that are safe, simple, non irritant and non-allergenic

- To use dressings that do not traumatize new tissues on removal and minimize frequency of changes

Physiology of Wound Healing

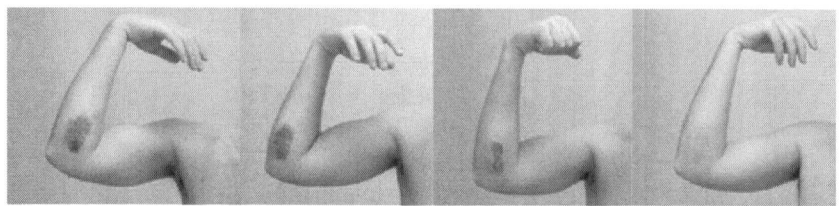

© Fedor Kondratenko Dreamtime.com

Wound healing is the term generally used to describe the mechanism through which the body repairs or replaces damaged tissue. An understanding of the healing process is essential if wounds are to be properly assessed and their management planned. Although the following headings are described separately, they often do overlap, particularly in chronic wounds.

This process can be described as progressing through three phases;
- Inflammatory Phase
- Proliferative Phase
- Maturation Phase

Wound healing varies considerably according to a number of factors which are discussed later. It is important to realize that although the surface of a wound may heal in days it might take months for the underlying tissue to heal properly.

The Inflammatory Phase (2-7 Days)

The first phase of healing is generally called inflammatory, because this is the most visible occurrence after injury. When tissue is damaged, blood vessels are injured and bleed into the space created. This blood then coagulates to form a fibrin clot, the injured blood vessels repair themselves, and the bleeding stops. The damaged tissue then secretes histamine, which has a number of effects, but principally acts to cause capillaries to constrict. During the next few hours, there is increasing tissue swelling and engorgement of surrounding blood vessels. This increased blood supply accounts for the inflamed appearance of; redness, warmth, swelling and Pain. Anything which prolongs this inflammatory phase can delay healing, e.g. infection or physical damage caused during dressing changes.

The Proliferative Phase (8-24 Days)

During this phase, different types of cells arrive at the wound site to defend against bacteria, remove dead tissue and begin the repair process. In this phase the cells produce new strands of collagen which is the main constituent of skin, tendons, ligaments and scar tissue. The peak rate of production of collagen in a wound healing by primary intention is between the fifth and seventh day. It is therefore important to take Vitamin C at this stage as the body is unable to store it.

The longer the initial inflammation phase due to damage or infection the more likely scar tissue will be formed. As the proliferative phase proceeds further, there is a rapid increase in the tensile strength of the wound and the numbers of capillaries begin to decrease towards normal levels. Inflammation also decreases but the wound may remain red, raised and itchy.

The Maturation Phase (24 Days – 2 Year)

During this phase there is a progressive decrease in the blood supply to the scar tissue. The collagen fibers inside the wound move around to make the repair stronger. The dusky red appearance of tissue changes to pale white scar tissue. The strength of the wound, having increased rapidly in the first three weeks of healing will have regained only 50% of the normal tensile strength of a skin wound within the first six weeks, with the amount of collagen in the scar continuing to increase for several months. As it then gradually reduces more, so the scar flattens and softens. The tension within the wound causes the collagen to orientate itself at right angles to the wound margins in a three-dimensional "lacing" effect.

Epithelialisation

In addition to these four overlapping phases a further process Epithelialisation needs to be considered when the wound has been healing by secondary intention.(see below) This process takes place before maturation in such wounds. Following injury, epithelial cells at the edge of the wound migrate over the wound surface. These cells can only migrate over live tissue, and, therefore, if debris, blood clots or scar tissue are present, they have to migrate below (scabs). The best epithelialisation occurs in moist wound environment.

Wounds heal either by *Primary or Secondary* Intention

Healing by Primary Intention

Healing by primary intention can occur when the wound edges are either adjacent as In the case of an incision or brought together and closed with sutures, clips, glue or some other form of wound closure. The wound goes through the healing stages described above, but the minimal amounts of blood and exudates will have formed a natural barrier at the wound surface after forty -eight hours. Such a wound can be exposed after this time: a dressing should not be removed unless absolutely necessary within the first two days.

Healing by Secondary Intention

Healing by secondary intention takes place when there has been tissue loss (often extensive) and there may be debris and exudates to be cleared from the wound. Granulation tissue will form the base and sides of the wound, increasing in thickness slowly filling the wound.

Factors which affect wound healing

The human body is able in favourable conditions to heal most wounds without any outside intervention. However a number of factors may contribute to wound break-down and influence wound healing.

Gender

Males shed bacteria more actively and have a higher chance of infection and subsequently reduced healing rate than females.

Infection

Infection of a wound occurs when the germs are sufficiently virulent to overcome the body's resistance Bacteria can affect the production of collagen and delay epithelialisation.

Haematoma

A build-up of blood within the tissues appears to have a toxic effect on the tissues as well as acting as a focus for infection. Also the pressures created can restrict blood supply and cause tissue damage.

Temperature of Wound

The best temperature for wound healing has been shown to be 37 degrees Celsius. Certainly extremes of heat and cold can show significant tissue damage. Hence, you should use warmed saline for irrigation/cleansing; and the avoid removing dressings unnecessarily

Nutritional Status

It is important to realize that malnourishment will slow healing: in particular the reduction of collagen synthesis as a result of protein deficiency. Deficiencies in Ascorbic Acid (Vitamin C, which contributes to collagen formation), zinc (probably through its effects on the enzyme systems involved in protein metabolism) and iron (through its effect on haemoglobin levels) have all been cited as factors affecting the rate of tissue healing. So it's worth stocking these supplements for medicinal use if none other.

Drugs

Although not usually possible to stop taking them if you have other problems its should be noted that the following drugs may have an adverse effect on wound healing;

Indomethacin (an *Anti-inflammatory)*, Aspirin, Cytotoxic Drugs and Steroids

Embedded Object in Wound

The presence of an Embedded Object in a wound can act as a focus for infection in addition to their ability to provoke an immune reaction.

Good Blood Supply

A wound is dependent on the blood for a supply of nutrients and cellular material vital for wound healing. Well supplied areas such as the face usually heal quickly whilst extremities of the limbs receive a poorer supply.

Smoking

Studies have demonstrated that blood flow to peripheral vessels is reduced by the inhalation of nicotine. Some research also points to direct tissue damage and inhibition of the healing process.

Illness

Anaemia may in itself delay healing due to the supply of oxygen to the wound being reduced. A Circulatory disease will, in limiting the flow of blood, have a detrimental effect on the rate of healing. Other Diseases that affect healing are Diabetes, Jaundice and Uraemia (increased Urea levels). Respiratory & Cardiac Disease`s (Reduces Oxygen to Tissues).

Patient`s Age

Ageing results in reduced skin elasticity and metabolic rate; there may be some muscle atrophy. Meanwhile, with increasing age, the inflammatory and repair mechanisms operate less vigorously and cells, therefore, are replaced more slowly. General blood supply may also be impaired.

Age of the Wound

It is probable that the older the wound when appropriate intervention commences, the slower healing will be. (So treat early)

Surgical Technique

Rough handling, can damage tissue and provide a focus for infection. Poor suturing technique can also influence wound healing since applied too tightly they can cut through a wound edge, whilst incorrectly inserted they can damage weakened tissue.

Remember that sutures are foreign bodies and although sterile can both evoke an immune response or drag contamination though a wound.

Types of Wounds

Necrotic Wounds

Necrotic tissue is dead tissue found in or around a wound. It is usually recognizable by its black or yellowish-brown colour. The first aim of managing such a wound is to remove the necrotic tissue without damaging the underlying or surrounding tissues.

Sloughy Wound

Slough is formed when dead cells accumulate in the exudates of a wound. Slough tends to be yellow in colour but care must be taken not to misinterpret slough either for, occasionally, epithelial tissue, or for the pus produced when an infection is present: the other signs and symptoms present which are likely to co-exist in an infected wound are outlined below. Further, some dressings may interact with the wound to produce fluids, which can be mistaken for slough or pus.

Granulating Wounds

Granulation tissue is recognised by its red 'healthy' appearance, as new blood vessels are formed and grow into the wound area. However, healthy granulating tissue, being well supplied with blood, bleeds easily and the avoidance of trauma and the promotion of granulation by providing a clean moist environment, are the aims of management at this stage, (Care must be taken however, not to assume a red wound equals a healthy wound. Infected wounds can also have a deep red coloration to them and the presence of any of the other signs of infection mentioned above must be taken seriously.

Granulation tissue

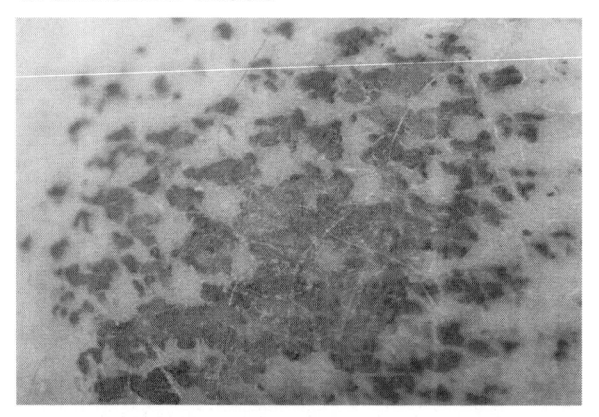

© Spe Dreamtime.com

Infected Wounds

Infected wounds often contain pus (frequently yellow in colour) and are likely to show signs of tissue destruction and delayed healing. There is likely to be some or all of the following signs and symptoms present: High temperature, heat and swelling around wound edge, inflammation, Cellulitis, offensive odour and greenish slough. Management must be aimed at treating the infection with antibiotics such as Flucloxacillin and/or providing the optimum environment for the body defence to overcome the invading infection.

Malodorous Wounds

Malodorous wounds have an offensive odour often arises when wound healing is complicated by anaerobic bacterial infection. Such wounds are often discharging exudates, lesions, as well as infected pressure sores and leg ulcers. Management of such wounds will often be dominated by the need to promote the patient's quality of life: care may, therefore on occasion be unconventional and aimed at symptom control rather than active treatment.

Epithelialising Wounds

Epithelial tissue begins to develop when a wound is filled with granulation tissue to the level of the surrounding skin. The colour of this type of wound can vary from pinkish through to yellowish and may require careful assessment to distinguish the epithelial tissue from slough. Fragile epithelial tissue develops best under moist conditions when cells can migrate readily across the surface of the wound: although the avoidance of trauma during dressing changes is an equally important requirement of management regime.

Sinus or Fistula

Almost any type of wound could be complicated by a sinus or fistula. The presence of pus might alert one to the possibility of a sinus within a wound, the pus being discharged into the wound cavity from the abscess beyond: whilst the presence of body fluid e.g. faecal fluid, urine, bile discharging either into a wound cavity or via a suture line suggests the presence of a fistula.

A sinus is often difficult to heal as the cavity beyond often contains foreign material e.g. suture material. The management of such a wound ideally requires that any foreign material is removed. The sinus must be opened sufficiently wide to allow any exudates to drain or it should be opened out (often surgically) and allowed to heal by granulation from below. The exudates from a draining sinus should be absorbed and a means of ensuring that its opening cannot heal must be adopted but a tight pack will act as a bung and prevent the first requirement of sinus management -the free drainage of exudates.

A fistula may develop as a result of a disease process or due to breakdown of a surgical repair. Fistulae often close spontaneously or (further) surgery may be required. In the interim such wounds are usually managed symptomatically with care of the skin to prevent excoriation being a major consideration.

A wound drainage system may be used for containment and collection of the discharging fluid - this also allows for an estimate of fluid loss to be made. Nutritional support will also be an important consideration in order to maximise the chances of normal healing. This will need particular consideration if it is a faecal fistula that is present.

Treating Traumatic Wounds

A number of specialized products are mentioned below, I don't recommend you stock these as you are hopefully unlikely to ever need them, some are POMs and they are expensive, but are worth bearing in mind for supply's. But even using normal first aid products the principle remains.

Wound Cleansing

Traumatic dirty wounds may be contaminated by dirt or bacteria so thorough cleansing of the wound and surrounding area is required. Irrigation of the wound by tap water or other cleaning fluid is the best way to remove contamination, in the field the best way is to use a large syringe (or possibly a water pistol/ Squeezable bottle). This should be followed a by gently scrubbing motion using a sterile sponge, gauze or wound tissue. More extensive cleansing to remove grit may be required and carried out under local anaesthetic.

If you have it use sterile saline, boiled then cooled water or chlorhexidine gluconate to reduce bacterial contamination. However it's a thorough cleaning action not the liquid that reduces the chance of infection

Wounds healing by secondary intention may require excessive discharge and debris to be cleared. Ongoing cleaning of a wound during showering (preferably), or bathing, avoids the trauma of formally cleansing the wound, where fibres of gauze or cotton wool may get stuck in it. Also a long healing wound does not need to be cleaned using a sterile technique and warm tap water is fine.

It follows then, that if mechanical cleansing of a wound is required it should be done by irrigation using a warmed 0.9% sodium chloride solution. If wetted swabs must be used to aid mechanical removal of debris etc. they are likely to cause less trauma if handled wearing sterile gloves rather than with forceps.

Lacerations

A laceration is a straight cut to the skin, often superficial but may involve deeper structures. The main aim of managing these wounds is to close the skin

For deep lacerations Providone iodine sprayed into the wound before closing protects against infection. A low-adherent absorbent dressing such as polyurethane foam is ideal with pressure bandaging added if there is much discharge after suturing.

Discharge is usually light and within forty-eight hours, a natural barrier against pathogenic invasion will have been formed and the wound can, if required, safely be exposed at this point. Continued cleaning of such wounds will be unnecessary unless there are signs of breakdown or infection

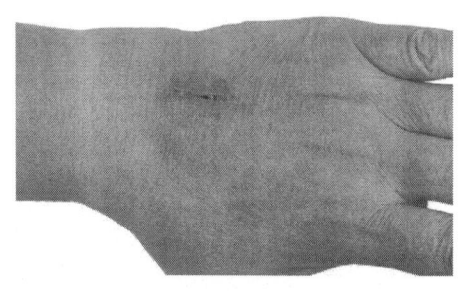

© Birgit Reitz-hofmann - Dreamtime.com

Crush Injuries

When dealing with Crushed fingers and toes you may need to fix an underlying fracture.

Often blood collects under the nail and is very painful, this is easily relieved by heating a paperclip and using it to melt a hole in he nail letting out the blood. If there is loss of a lot of tissue they should be dealt with as open wounds. A low-adherent absorbent dressing such as polyurethane foam or an alginate sheet should be used.

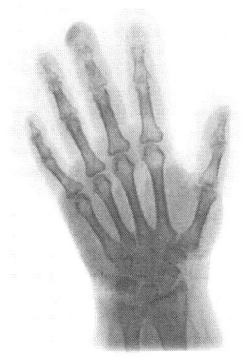

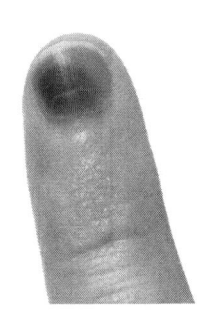

© Tallik | Dreamstime.com   © Marcin Pawinski | Dreamstime.com

Deep Penetrating Wounds

Penetrating wounds consist of a narrow but deeply penetrating tract; they may contain infection buried deep in the wound. Closing the wound is not an option as it will trap the infection. These wounds need to be opened to more under anaesthetic to be cleaned properly. Following cleaning and depending on the size of the hole, either a biodegradable alginate ribbon or a hydrogel with ribbon gauze, loosely packed, are recommended.

Abrasions

Abraisions, although apparently superficial are usually very painful, and commonly have dirt and grit in the wound. They must be thoroughly cleaned and the dressing used must prevent further contamination and absorb discharge. Following thorough cleansing, a semi permeable film dressing can be used for low discharge wounds whilst a polyurethane foam or hydrocolloid sheet is recommended for a wound with high discharge. For high risk wounds use an antiseptic-impregnated dressing

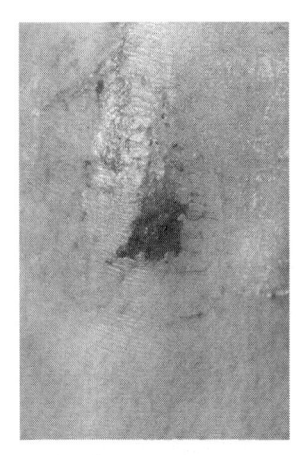

© Szefei | Dreamstime.com

Bites

Bite injuries produce a confined ragged wound with a high risk of infection. Thorough cleaning is essential. Bites should not be closed A short course (5 days) of antibiotics is usually advisable (e.g. Flucloxacillin, Augmentin or Cephradine)

De-gloving Injuries

De-gloving wounds cause layers of tissue to be torn away exposing deep tissues and may cause extensive damage to skin, fat muscle and bone. Such injuries will need thorough cleaning and possible skin grafting.

Use of Antibiotics

Most wounds will contain some bacteria, and chronic wounds may be heavily contaminated if there is debris or dead (necrotic) tissue. We don't have the benefit of Laboratory test so need to use antibiotics where contamination is likely or signs of infection are present.

If infection exists, appropriate oral antibiotics should be started. Oral antibiotics will also be used following bite injuries. The only indication for the use of topical antibiotic is for an infected/malodorous wound when topical Metronidazole may be used.

Dressings

If a wound is infected or smelly the dressing should be changed daily, otherwise they can be left in place for up to 7 days, in all cases follow the manufacture instructions.

Selection of Dressings

For more details of Wound Dressings please refer to a recent edition of the British National Formulary [BNF].

Necrotic/Sloughy Wounds

Exudate	Infected	Cavity	Choice 1	Choice 2
High	No	No	Alginate/Hydrofibre	Foam
High	No	Yes	Alginate/Hydrofibre	Foam
High	Yes	No	Alginate/Hydrofibre	Foam
High	Yes	Yes	Alginate/Hydrofibre	Sugar Paste
Low	No	No	Hydrocolloid	Sugar Paste
Low	No	Yes	Hydrocolloid	Sugar Paste
Low	Yes	No	Hydrocolloid	Sugar Paste
Low	Yes	Yes	Sugar Paste	Hydrogel

Granulating Wounds

Exudate	Infected	Cavity	Choice 1	Choice 2
High	No	No	Alginate/Hydrofibre	Foam
High	No	Yes	Alginate/Hydrofibre	Foam
High	Yes	No	Alginate/Hydrofibre	Foam
High	Yes	Yes	Alginate/Hydrofibre	Foam
Low	No	No	Foam	Hydrocolloid
Low	No	Yes	Foam	Hydrocolloid
Low	Yes	No	Foam	Hydrocolloid
Low	Yes	Yes	Foam	Hydrogel/ metronidazole Gel

Epithelialising Wounds

Exudate	Infected	Cavity	Choice 1	Choice 2
High	No	No	Foam	Non-adherent
High	No	Yes	Hydrogel	Cavity Foam
High	Yes	No	Alginate/Hydrofibre	Foam
High	Yes	Yes	Alginate/Hydrofibre	Sugar paste
Low	No	No	Foam	Non-adherent
Low	No	Yes	Hydrogel	Cavity Foam
Low	Yes	No	Foam	Hydrocolloid
Low	Yes	Yes	Sugar paste	Hydrogel

Wound Closure

Wounds heal on their own, the purpose of wound closure is to realign damaged tissues to restore its original function and strength.

Primary Closure

In Primary Closure the wound is closed immediately using one of the methods below

Secondary Closure

In Secondary Closure the wound cannot be closed immediately and is left for granulation to take place

Tertiary Closure

In Tertiary Closure the wound carries a high infection risk and needs to be left open whilst oral antibiotics and wound packs fight the infection. The edges of the wound then need debridement to encourage healing and the wound is closed.

Methods of Closure

There are four methods of commonly used wound closure;

Steri-strips

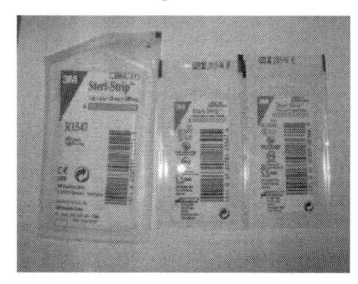

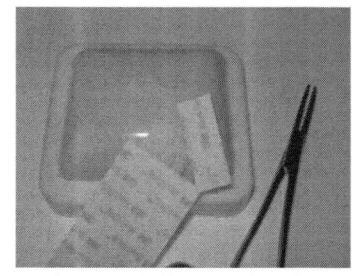

Steri-strips are porous adhesive paper strips use to close wounds. They have several advantages over other methods;

- Cheap to use
- Quick & Easy to apply
- Painless
- Don't require specialist removal

There are some disadvantages;

- Can't easily used on a joint as movement will loosen them.
- Can't get them wet
- Can't be used on hairy areas

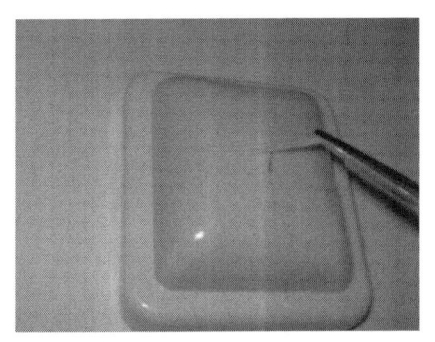

When applying steri-strips place one across the centre point of the wound. Then work from the centre placing alternative strips each side the centre one at 3mm intervals. The gaps allows exudates to escape from the wound. If the wound is particularly wide then reinforcement strips can be placed across others parallel to the wound.

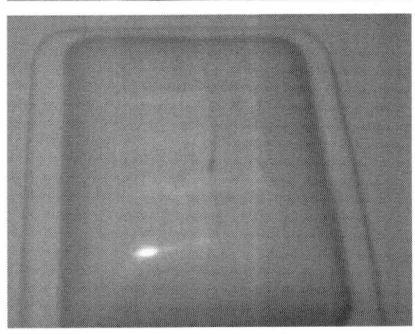

If covering with a dressing, ensure it covers the area completely to prevent infection, keep dry and secure strips.

Wound / Skin Glue

Skin glue is a sterile version derived from super glue. Make sure bleeding has stopped and the wound is clean and dry before applying glue. It's useful for small wounds and is applied either in spots or as a single line. The glue hardens in around 30 seconds and the edges of the wound need to be kept together during that time. The area will need to be kept dry for at least 5 days. Commercial glues include Dermabond, Histoacryl and epiglue. Prices vary between £5-£10 a single use tube.

Skin Staples

© Christian Delbert | Dreamstime.com

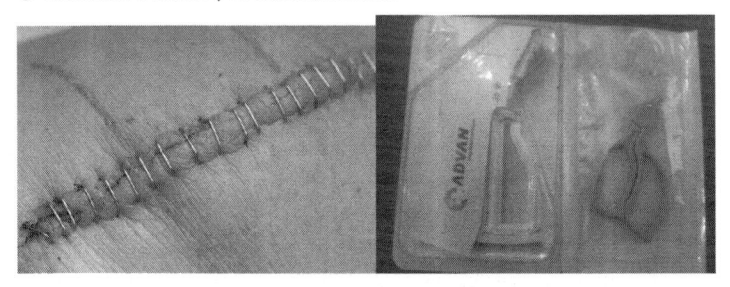

Skin staples are useful for closing minor wounds, they are easy to apply not requiring the level of skill required for suturing and are more durable than steristrips. They can be used in areas where body hair would need to be shaved for steristrips and can get wet unlike steristrips. When applying press the gun firmly against the skin. The staples cause momentary pain when going in but applying local anaesthetic first will probably hurt as much. A special tool is required to remove the staples for minor wounds they can be removed after 5 days.

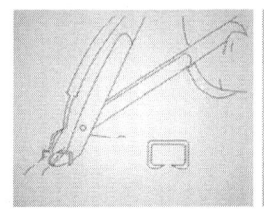

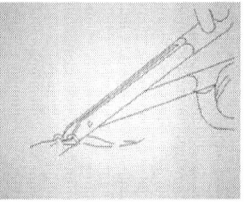

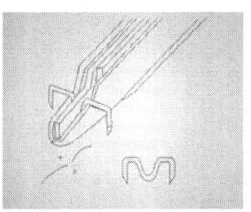

Sutures

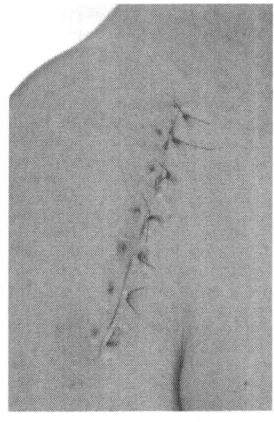

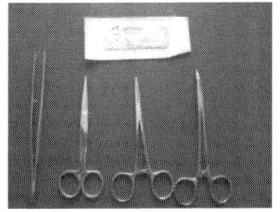

Sutures in the past have been considered the best way to close a wound although they have several disadvantages other methods should be considered first.

They can be used to close layer of tissue below the skin if the wound is gaping. Some disadvantages are; that a level of skill is required to suture, specialist equipment is required, it further damages the skin at suture points and is painful for the patient.

© Birgit Reitz-hofmann | Dreamstime.com

For a larger wound local anaesthetic may be required to both clean and close a wound. This is covered later but will require a local anaesthetic agent, wound pack, syringes and needles, needle holder and forceps. All equipment should be sterile and the practitioner should use sterile gloves.

Firstly clean wound using sterile technique. Draw up local anaesthetic into syringe and attach needle, See section on local anaesthetic for administration Open packets and place instruments on sterile sheet (field) from wound pack without touching them, put on sterile gloves.

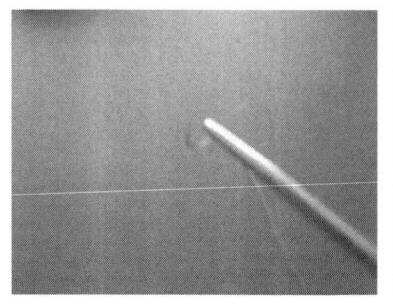

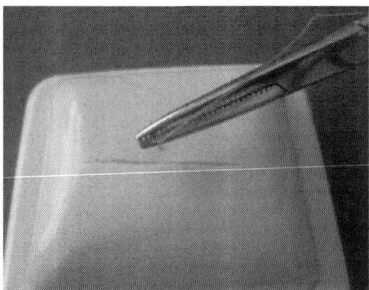

Use needle holder to hold needle at end in a comfortable position. Start suturing in the middle of a wound then move outwards in the same manner as for steristrips. Make sure the tissue is lined up correctly.

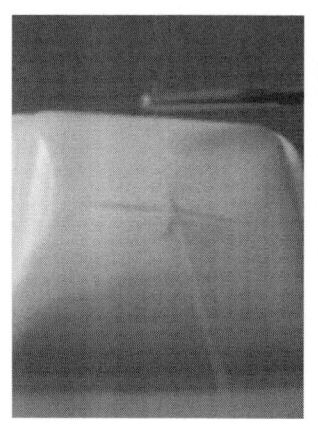

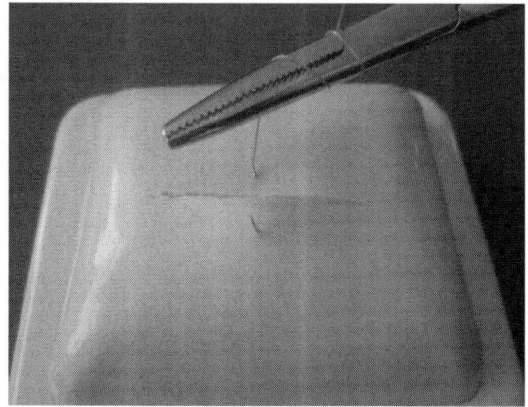

The suture needs to go deep enough to pull layers of tissue together and not leave a gap under the skin where blood and infection may gather.

Place suture needle in holder, use forceps to handle tissue gently, pierce skin with needle taking a 'bite' of tissue, pass needle though opposing side of wound and back up through the skin.

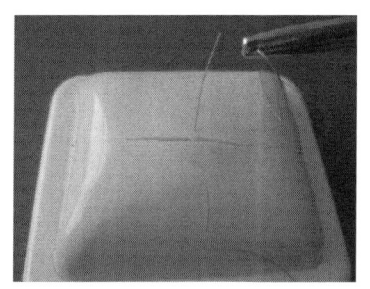

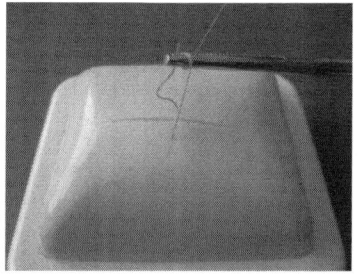

Pull suture through leaving enough to tie off.

Loop thread twice clockwise around forceps, take loose end in jaws and pull knot tight then secure with a second single knot by looping anti-clockwise once around the forceps.

Repeat first double clockwise loop to fully secure it. Don`t tie knot over wound a this may irritate it and keep to all knots to one side trim ends of suture.

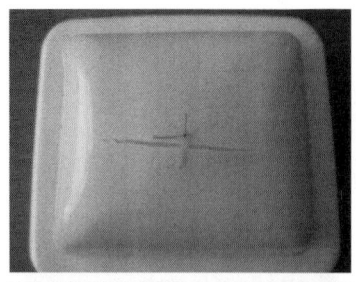

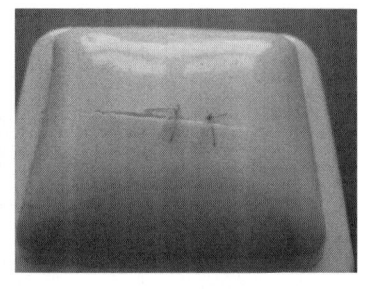

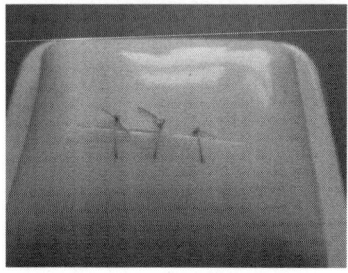

Move on to next suture, placing alternative sutures each side the middle one until wound is closed and stable.

Sutures come in a large variety of sizes and materials, Some are absorbable and are used to close deeper layer of tissues or in surgical repair.

Most sutures come with needles and thread attached together in a single packet, however you can get them separately and get packs with multiple sutures in them.

The sizes run from largest to smallest as 7,6,5,4,3,2,1,0,2-0,3-0,4-0,5-0 to 11-0

Actual diameter of thread depends on material, but as an example for non-absorbable suture a size 11-0 would be 0.01mm in diameter and a size 6 would be 0.8nn in diameter.

Needle shapes and tips also vary but a curved needle is generally used for wound closure.

A stitch cutter is used to remove the stitches which is similar to a scalpel blade. The time stitches need to remain in situ will vary for example;

- Facial 3-5 days
- Scalp 7-10 days
- Limbs 10-14 days
- Joints 14 days
- Torso 7-10 days

Suturing is not a skill you want to practice for the first time on a patient in the field. If you intend to carry the equipment practice before you need to do it for real. There are a number of good tutorial videos and guides available on the internet. You can practice on belly pork, oranges, foam packing and can get practice pads intended for injections as shown above.

Treatment of Burns

Burns are fourth major cause of trauma related deaths, Burns can be classified into three types;

- Superficial (First Degree)
- Partial thickness (Second Degree)
- Full thickness (Third Degree)

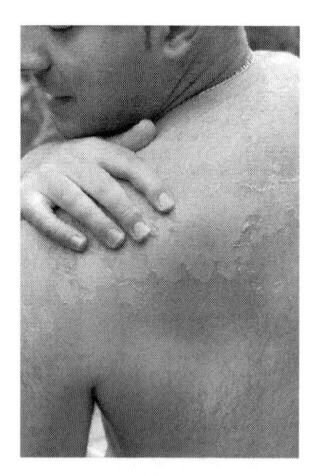

© Arenacreative | Dreamstime.com

Superficial burns only affect the top layer of skin, they are denoted by reddening and swelling of skin and tenderness.

Partial thickness burns affect the epidermis causing reddening and rawness. Blisters are formed from plasma released from tissues.

Full thickness burns affect multiple layers of skin and can affect nerves, blood vessels and underlying muscles.

A casualty who is trapped in a burning structure or vehicle can experience a number of problems apart from tissue lose. As always our priority is to maintain (A)irway, (B)reathing and (C)irculation. A patient who inhales superheated air or steam is going to develop airway problems as the soft tissues in their upper airway swells reducing their ability to effectively ventilate and provide oxygenation. It's Important therefore to assess their respirations for adequacy and depth. This is a situation where endotracheal intubation is required over a Laryngeal Mask Airway (LMA) or other simple adjunct for the unconscious patient, before the airway closes due to swelling. In all cases the patient will benefit from oxygen and may require assisted ventilation with a Bag valve and mask.

If intubation is not performed early a cricothyrotomy or tracheotomy may be needed to maintain a patent airway. Toxic inhalation may occur from carbon monoxide if petrol, wood or oil was burnt in the fire or from cyanide poisoning if nylon or polyurethane was burned.

Severe industrial Cyanide poisoning is treated with 300ml dicobalt edetate iv followed by 250ml of 10% glucose iv. Premixed antidote kits are available from laboratories that routinely use cyanide.

All casualties that receive severe burns will benefit from iv fluid therapy to replace lost liquid from plasma. The amount of fluid required is often underestimated, first we need to calculate the body surface area[BSA] that has partial or full thickness burns. This can be done using the Rule of 9s.

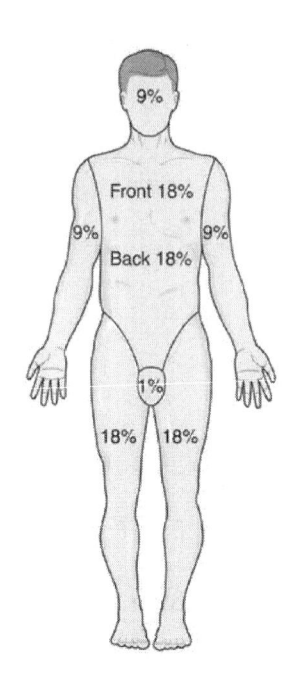

The Rule of 9s.

Head [9%]
Chest and Abdomen [18%]
Back [18%]
Each arm [9%
Each leg [18%]
Genitals [1%]

Initial fluid requirement should be based on 20ml/kg weight of the patient up to 2 litre`s. Fluid required in the first 24 hours is often underestimated and is based on;

4ml x patient weight in kg x BSA burned.

With half the total being given in the first 8 hours.

Thus a 80kg man with 40% burns would initially need 1.6 liters of fluid, another 4.8 liters in the first 8hrs and a further 6.4 liters in the next 16 hours. For a total of 12.8 liters in the first 24hrs. If the patient is unconscious all of this would need to be given iv. Significant or critical burns are gauged as full thickness covering 10%+ of BSA or partial thickness of 30%+ of BSA or burns affecting the hands, feet, face, airway or genitalia. Probable mortality is calculated as the patients age + BSA burnt as a percentage.

Bizarrely one of the killers in burn victims is hypothermia, where blood plasma seeps into burnt areas is evaporated which leads to rapid heat loss and severe hypothermia.

Initial management of burns is cooling with clean cold water. Removal of restrictive items, smoldering clothing and application of saline soaked sterile dressings. Commercial dressings such as waterjel are available that contain lignocaine a local analgesia, however they should be used with caution in children as they can quickly lower their body temperatures, these are useful when dealing with superficial or partial thickness burns where nerves are still intact.

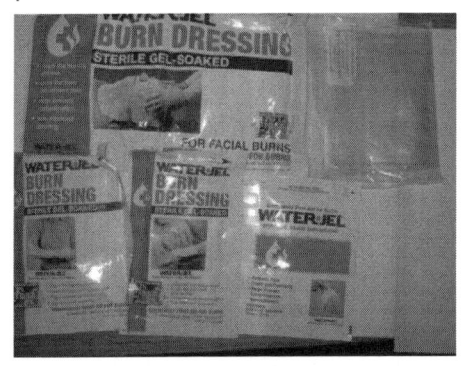

Cling film is also very good for dealing with burns it both stops oxygen from reaching the wound, preventing further heat damage and provides a sterile covering. When laying it over the wound place it in sheets rather than wrapping around a limb which if it swells will cause constriction.

For burnt hands sterile burn bags are available, if transit to hospital is delayed these can be filled with flamazine cream. A good natural treatment is tea tree soaked dressings.

Patients will also require some form of fast acting analgesia such as entonox or iv tramadol or morphine. Full thickness burns are often not painful as underlying nerves have been destroyed.

Burn creams and ointments should only be applied to superficial burns that do not require hospital treatment. When caring long term for burns victims, any blackened necrotic (dead) issue needs to be removed exposing pink healthy tissue below. It's important to keep the wounds moist to encourage healing, the downside of this is that this also increases the risk of bacterial infection so prophylactic antibiotic cover is essential, using Flucloxacillin either oral or by intramuscular 250-500mg injection QDS. Alternatives can be tried if resistant or allergic to Penicillin.

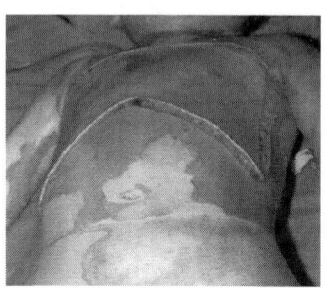

Here we can see the Burn skin has been cut around the chest to prevent it tightening and restricting breathing

Wounds that have significant skin loss may require grafting. Thin slices of skin are removed from donor areas. Skin grafts can also be synthetic or taken from cadavers. The graft is then cut to form a web of strands and is placed on the wound.

Local Anaesthetic

This has many uses in the wilderness setting as it allows the practitioner to deaden an area of the patients body which allows painful procedures to be carried out more comfortably.

It has advantages over general anaesthetic as the patient doesn't need to be paralysed or sedated.

Common uses for Local Anaesthetic are;

- Cleaning deep wounds
- Closing wounds
- Reducing Fractures
- Reducing Dislocations
- Dental Treatment
- Eye Treatment
- Pain Management
- Labour
- Orthopaedics
- Surgery

There are many methods of introducing local anaesthesia the ones we will consider are
- Topical anaesthesia
- Infiltration anaesthesia
- Nerve blocks

Topical anaesthesia

Any anaesthetic agent which is applied to the skin or membrane such as amethocaine eye drops (See Eye Injuries). Anaesthetic creams such as EMLA and ametop gel are sold in chemists for use when having tattoos and piercings they are used medically for numbing skin whilst taking blood or inserting cannulas, lidocaine throat sprays are also available at chemists who treatment of sore throats.

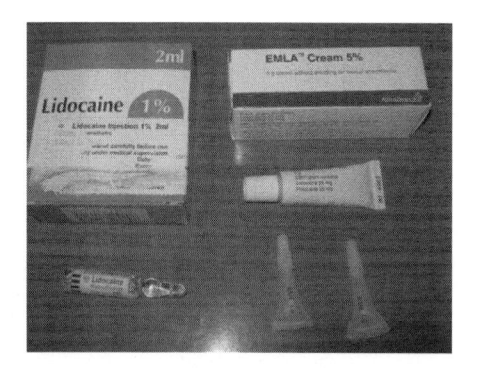

Infiltration anaesthesia

Injection into tissue used in suturing and wound cleaning. A number of local anaesthetics are available but Lidocaine is most commonly used. The maximum dose for infiltration is 200mg or 3mg/Kg whichever is less. Local anaesthetic increase the size of blood vessels (vasodilator) adding adrenaline (a vasoconstrictor) to Lidocaine at a concentration of 1 in 200,000 slows rate of absorption and prolongs its local effect.

Lidocaine for injection comes in three strengths 0.5%(5mg/ml), 1%(10mg/ml), 2%(20mg/ml) and as xylocaine 1% with adrenaline.

Due to its vasoconstrictor effect xylocaine with adrenaline must not be used on fingers, ties, noses and other extremities as it may cause tissue necrosis (death).

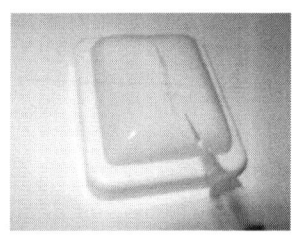

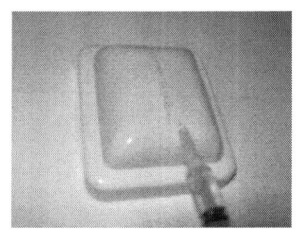

The best place for introducing anaesthetic is just under the skin through the side of a wound, this prevents additional punctures damage to the skin. However if the wound is dirty this may push contamination further into the tissue. An alternative method is to inject through four points in a diamond pattern around the wound.

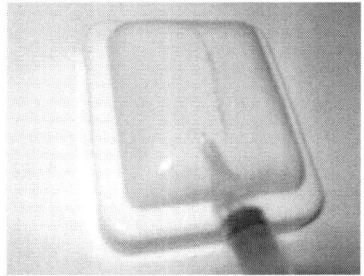

Whichever method is used draw anaesthetic up into a syringe with fine needle such as 23 or 25 gauge. Insert needle to full depth along the side of the wound then inject drug as you slowly withdraw it. If the wound is longer than the needle it can be reinserted at a point that is already numb

Nerve Blocks

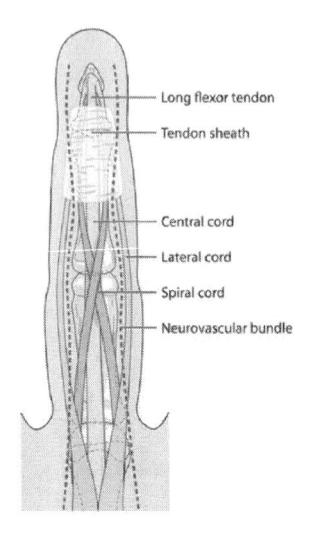

© Legger | Dreamstime.com

To aid procedures on finger and toes a digital nerve block can be used.

There are four nerves that run close to the bone two on each side one near the top and the other near the base of the finger

Three injection need to be made using a fine needle one each side on the bone, insert needle from top to bottom stopping before it penetrates the other side them inject drug as you withdraw, the final injection is off a small amount across the top of the finger.

Before you start injecting withdraw the plunger slightly to check you are not in blood vessel. If blood is aspirated the withdraw and try another spot. A further problem with this technique is that injecting too much fluid into a small space can compress blood vessels and thus cause ischemia and tissue death.

The total dose of 1% Lidocaine shouldn't exceed 4ml and takes 10 minutes to reach full effect. Check patient is numb before proceeding.

A similar technique can be used to numb the arm or leg using an axilla or femoral block but should only be attempted by a skilled practitioner.

Head Injuries

A head injury often causes nothing more than a bruise, lump or headache. There are however several serious conditions which need to be excluded.

Concussion

Concussion is where the brain, which is suspended in CSF shakes within the skull hitting its inside surface, the casualty may briefly lose consciousness and become confused. There may be a brief loss of memory, followed by nausea, dizziness and headache from a few hours to a few days. Recovery is usually complete.

Compression

Compression is usually caused by an injury due to either a skull fracture or laceration to brain where it has hit inside of skull due to sudden deceleration. Compression can be caused by bleeding inside the skull which holds a fixed volume of material or by swelling of brain tissue caused by trauma or infection. Pressure on brain tissue causes patients to display symptoms below as well as increasing disability.

In severe cases the brain can become herniated as it is pushed down into the opening where the spinal column joins the brain. This has the potential to be fatal.

Management

Any patient with a head injury who has a significant mechanism of injury, say from a RTC, fall or blow also has the potential for a spinal injury.

Assess level of consciousness using the Glasgow coma scale (see appendix 1)

A physical examination of the casualty should be performed. Look for blood with straw coloured fluid from the nose and ear this is cerebral spinal fluid (CSF) and is indicative or a basal skull fracture. Other signs are panda eyes, which are black circles of bruising around one or both eyes and battle sign which is bruising behind the ear.

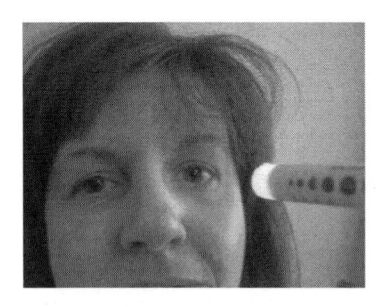

Check pupils reactions by shining a light directly into the casualties eyes, this is best done in dim light. Note the size of each pupil in millimetres, some pen torches have a gauge on the side showing pupils sizes.

The pupil should go smaller (constrict) in reaction to the light, note if a reaction takes place, if it is brisk or sluggish and if both eyes react the same. A difference in pupil sizes or reaction speed is often a clue to an underlying head injury. check to see if they can follow the light from side to side and if their vision is blurred.

Check the casualty can remember what happens (amnesia) a loss of less than a minute is rarely serious. Whereas a loss of greater than 30 minutes usually indicates a serious injury.

Note any history of unconsciousness although not a definite guide less than two minutes is associated with concussion whereas longer periods with compression which indicates bleeding or swelling off the brain

Further checks can include;

Hearing in both ears, if they can move all limbs, have normal sensation and no numbness or pins & needles. Finally check to see if they can stand and walk without staggering.

Even if the patient passes all the tests they should still be monitored for possible deterioration. As initial injury may cause swelling of brain tissue or slow bleeding into the brain. Signs they should be alert for is;

- Increased drowsiness
- Increased headache
- Confusion
- More than one episode of vomiting
- Weakness in Limb
- Facial Droop or Speech difficulties
- Dizziness, loss of balance
- Blurred vision
- Difficulty breathing
- Convulsions or Absences

It is difficult to manage head injuries outside hospital as those with serious head injuries usually are admitted to Neurological Intensive care unit.

However initial management includes;

Ensuring ABC and C Spine is protected

Record blood Pressure, pulse and respiration rate ever 15 minutes.

Maintaining blood pressure using patient positioning and intravenous fluids (max 1.5L/day) to ensure adequate oxygen reaches the brain. The figure to aim for is a Mean Arterial Pressure(MAP) >60mmHg, normal MAP is between 70-110mmHg this is determined by the following;

MAP= Diastolic BP+1/3(Systolic BP - Diastolic BP)

e.g a person with a Blood Pressure of 120/60

70+1/3(120-60) = 90mmHg MAP OK

e.g a person with a Blood Pressure of 60/30

30+1/3(60-30) = 40mmHg MAP too Low

Good pain control and treatment for nausea. Although you shouldn`t deny medication, don`t over medicate as this may mask signs of a deteriorating level of consciousness.

If chance of definitive care is delayed allow the patient to sleep as this will help the body deal with swelling (oedema) of the brain.

Allow the patient to lay down but elevate the head to 30 degrees, this will reduce headache and swelling.

Treat fitting with diazepam 10mg IV for adults. If swelling (oedema) is suspected treated cautiously with a diuretic such as mannitol 20% at 1g/kg iv over 10-20 minutes. This takes 20 minutes to take effect and lasts for 2-6 hours or Furosemide 0.4 mg/kg. Diuretics are most useful in the first 12-24 hours after that they can have a negative effect. They reduce swelling but also reduce blood pressure.

If consciousness reduced , hyperventilation with a bag, valve and mask at a rate of 20 breaths/min, this causes the blood vessels in the brain to shrink (vasoconstriction) and reduces intracranial pressure. If oxygen is available give 100%.

If there is an improvement within the first two days this is very good sign and usually indicates symptoms were from a mild concussion.

The risk of a clot is low in a casualty who is fully alert without a skull fracture (1:1000) but very high if confused with a skull fracture (1:4)

Chest Wounds

There are several different chest injuries.

Fractures of the chest wall have been previously covered.

Remaining injuries either effect;

- Lungs
- Diaphragm
- Trachea
- Heart
- Major Blood Vessels
- Oesophagus

Lungs

There are a number of lung injuries that can result from trauma;

Pulmonary contusion

Pulmonary contusion is bruising of the lungs causing bleeding into the alveoli. This decreases gaseous exchange and lowers oxygen level in blood. Support patient with IV fluids and oxygen if available.

Simple Pneumothorax

A simple pneumothorax is caused by trauma or can happen spontaneously. Degree of compromise to breathing depends on size of pneumothorax. Small tears will heal themselves, monitor in case it become a tension pneumothorax and treat symptomatically.

Open Pneumothorax

An open pneumothorax is known as a 'sucking chest wound' caused by penetrating trauma which will allow air to enter the chest cavity from outside. Bubbling of blood in the wound may be seen on expiration. As may subcutaneous emphysema occur which is air in the surrounding tissue. This give a 'bubble wrap' feel to them.

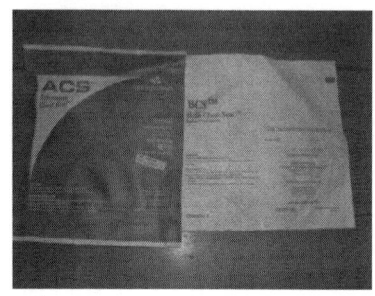

If available use a asherman chest seal or bolin chest seal to cover the wound this will allow air out of the chest but prevent more being sucked in. If there is more than one wound place occlusive dressings over others.

Tension Pneumothorax

A Tension Pneumothorax occurs when air enters the pleura but cannot escape, as pressure builds up it collapses the lung on the effected side, as it progresses it can also affect the heart and other lung. Potential signs include; diminished or absent breath sound, unilateral chest movement, fast breathing, lowering blood pressure, distended neck veins and cyanosis. If you see the trachea has moved away from the midline towards the unaffected side this is a very serious problem. Support blood pressure and oxygenation. The definitive treatment is by needle decompression.

Locate point of entry between 2nd and 3rd ribs in a line traced down from the middle of the collar bone. This is known as 2nd intercostals space, midclavicular line.

Insert a needle (max length 3.25") over the top border of the 3rd rib at 90 degrees into the chest. This position will keep it away from nerves and blood vessels that run along the bottom of ribs and from puncturing the heart.

You may feel a pop or hear a hiss of air as the needle releases the pressure. A good tip is to attach a syringe to the needle this will provide stability and you may see the plunger rise as air is released from the chest.

Once decompression has been achieved stabilise the needle in position and cover with a chest seal if available.

Heamothorax

A heamothorax is bleeding into the chest cavity will present with both breathing and circulation problems. Which will require support for both. A chest drain is the definitive treatment when surgery is not available,

Kits are available with all the necessary equipment to perform the procedure, the outer tissue is cut with the scalpel then the underlying tissues are parted by blunt dissection. If cut with a knife the tissues will take longer to heal. The tube should be placed in the sixth or seventh rib (intercostals) space at the posterior axillary (rear of armpit) line.

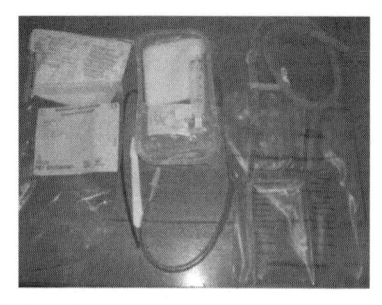

Diaphragm

This is a sheet of muscle that stretches across the bottom of the chest and is an essential part of respiration. It can be damaged by blunt or penetrating trauma, it can also be damaged by pressure changes during an explosion. If it tears abdominal organs may become herniated through the tear this will interfere with respiration and may damage the organs. It requires surgery to repair.

Trachea

Damage to the trachea will affect breathing. Monitor respiration rate, oxygen saturation, movement of chest, listen for unequal or absent breath sounds. Give oxygen and suction airway, avoid intubation and positive ventilation with bag, valve and mask unless absolutely necessary as it may further damage the airway. Cover any open wounds, infection is a common problem so prophylactic antibiotics should be given.

Heart

Two conditions affect the heart during trauma;

Myocardial Contusions which is bruising of the heart muscle, this shows the same signs as a heart attack and can cause cardiogenic shock it should be treated in the same way with careful monitoring of patient vital signs particularly blood pressure and oxygen saturations. It presents as central chest pain following a history of chest trauma.

Cardiac Tamponade which is bleeding inside the sac surrounding the heart. As the blood builds up it compresses the ventricles of the heart and rapidly affect the heart function. It presents with three symptoms known collectively as becks triad; a low blood pressure, muffled heart sounds and distended neck veins. They are often also breathless and confused. IV fluids should be given to maintain a radial pulse. There is a technique called pericardiocentesis. Which involves inserting a needle with syringe though the chest wall into the pericardial sac and removing blood, as little as 30-50mL may produce dramatic hemodynamic improvement. However doing so without x-ray equipment is extremely hazardous.

Major Blood Vessels

Vessels can be damaged by blunt or penetrating trauma, explosion or deceleration injuries. If the aorta is completely severed death is rapid, if it is damaged but not severed a third of casualties die in first 24 hours and half within 48 hours. Aortic dissection is characterised by sudden onset ripping chest pain. Pain is often said to go through to back or into neck or jaw. Blood pressure is often low. Although major damage requires surgery. Smaller tears may be managed medically by reducing heart rate (60-80 bpm) and systolic blood pressure (100-120) with beta blockers.

Oesophagus

In traumatic injuries the Oesophagus can be damaged by penetrating trauma. The main problem with this is that it allows gastric content to enter the chest cavity. This can lead to pneumonia and sepsis. Treatment should include;

- Oxygen
- Keep patient nil by mouth
- Insert Naso-gastric (NG) tube to remove stomach contents
- Give IV Fluids
- IV Antibiotics
- IV Anti-emetics & IV Analgesia

Small tears may heal themselves, larger ones require surgery.

Abdominal Injuries

The abdomen contains many organs, it is separated from the thoracic cavity by the diaphragm. If the injury occurs before inspiration then the diaphragm will be high in the chest, at around nipple level. If the chest is fully expanded then the diaphragm will be flattened. This has to be considered when evaluating injuries. Three types of structure exist in the abdomen;

- Solid Organs
- Hollow Organs
- Vascular Structures

Solid Organs include the Liver, Kidney, Pancreas and Spleen these tend to be very vascular and damage causes serious bleeding. Hollow Organs such as Intestines, Stomach, bladder and Gallbladder these rupture when damaged causing they contents to spill into the abdomen, leading to severe infection. The third group of structures include the Aorta, Femoral Artery and Vena Cava, if these are damaged it usually causes life threatening bleeding. If the diaphragm ruptures then abdominal content will enter the thoracic cavity become herniated and interfere with breathing.

Damage to the abdomen may be caused by Blunt, Penetrating, Shearing, deceleration and Blast injuries. If the wound is open then the organs may protrude from the abdomen. These should be covered with damp dressings or sheets of cling films, do not allow them to dry out.

Managing Road Traffic Collisions (RTCs)

Your ability to treat casualties at an accident is both limited by your expertise and equipment that is available.

A few general principle apply to most road traffic collisions. If you stop at an accident do so behind the incident blocking the road. Park you vehicle 'defensively' at an angle rather than straight in the road. that way if traffic behind you fails to spot the accident and unfortunately hits your vehicle it won`t be pushed into the accident causing further injury to the casualties or yourself.

Assess the scene for possible hazards such as fire, fuel spills and hazardous loads. Only approach the scene if its safe to do so. If it is dark and you have one wear a Hi-viz vest or coat.

Call the police and ambulance if it's possible to do so.

If you are not alone send another person to set up a warning triangle between 100-200 meters behind you vehicle and if it is dark break some snap light on the road. Both should be carried in your vehicle together with a adequate first aid kit.

If possible approach any vehicle occupant from the front, call to them say you are there to help and tell them not to move. If they do have a spinal injury the last thing you want them to do is look over their shoulder to see who is coming behind them. the emergency services bang the bonnet of the car to focused the casualties attention forward whilst talking to them.

If the car doors are locked or jammed you may need to gain entry my smashing a window a rescue hammer is good for this and usually comes with a seatbelt cutter as well. A thick pair of gloves is essential too.

Ensure you minimise further danger. ALL engines to be switched off, Don't allow people to smoke nearby.

Check that casualties are breathing, if not you need to get them out of the car as quickly as possible to perform CPR, it's very difficult to do this when someone is in a vehicle and should only be attempted insitu if they are trapped.

If the casualty is conscious and has suffered any type of head injury or was hit from any side with force, always consider the possibility of a spinal injury and try and encourage them to stay in the car if it's safe to do so.

If possible get them or help them to sit upright and support their head/neck from behind in a neutral position, don`t flex the neck if you can help it. You may need to kneel on the backseat of their car to do this effectively, always explain why you are doing things to the casualty.

If they are unconscious but breathing inset an oral or nasal airway(if you have them) and open there airway using the Jaw Thrust method if you are trained to do.

If the occupants are conscious check that they are all accounted for, there has been many cases where unrestrained children and adults have been ejected from vehicles Into ditches or over hedges and not immediately been found and of babies and small children that were on peoples laps and have been thrown into the foot well. or have been injured wandered away from the accident then collapsed.

If there are casualties on the road who are not conscious try to get someone to stay with them to monitor here breathing but don`t move them unless you have too. if you don`t have enough people, try and get them into the recovery position whilst supporting there neck as best as you can to stop it moving about. Treat any bleeding and support any fractures before moving them.

The decision as to whether to remove someone from a vehicle can sometimes be difficult to make particularly if you don`t have the equipment or personnel to move them safely.

If they are not breathing or are in imminent danger then they need to come out quickly. If you think there condition is deteriorating such as changes in breathing rate, pulse or Level of consciousness you should try and get them out in a control manner supporting their spine, airway and any injuries they have sustained. If you do have to move them place them on their back in a neutral position and don`t leave them.

Other dangers may come from traffic or environmental factors where in your judgement leaving them where they are is more dangerous than moving them. In these situations you need to use common sense.

I`ve mentioned 'Neutral Position' several time and this may need further explanation. Its keeping the head upright and in line with the spine, think of it as soldier at attention.

If equipping a vehicle you may want to consider

stocking it with the required Basic equipment; Hi-Viz Vest(£8), Rescue Hammer/Seatbelt cutter (£8), Small Fire extinguisher (£10), Heavy work gloves, Warning Triangle(£8), Snap Lights(£1 ea), Torch, Standard First Aid Kit(£10+).

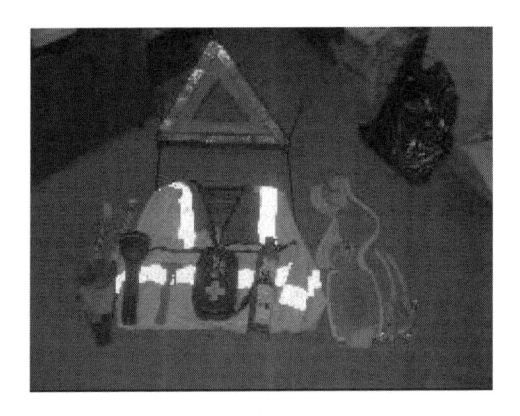

Some additional equipment which may be worth considering without going to overboard are , Oral Airways (About 50p each or £2.50 a set of 6), Nasal Airways (£4 each) or Stiff Neck Extrication Collar (between £6-£13ea). Two small PMRs (Radios) are useful allow those directing traffic to communicate and also for personal security. A bright flashing beacon will gain attention far more than warning triangles alone when placed on the road or a vehicle.

Chapter 7 Joint Problems

Arthritis

Any inflammation of a joint is known as arthritis. There are different type of arthritis.

Rheumatoid arthritis

This is a common form of arthritis affecting about 1% of the population, three quarter of sufferers being women. In these people the body makes antibodies that attack the synovium causing inflammation, this can damage the joint, cartilage and adjacent bones.

It can affect a few or many joints. Most often small joints are affected such as fingers, wrists, ankles and knees, although any joint can be affected. Rheumatoid arthritis is a chronic disease, the severity and time between episodes is different for each patient.

Symptoms include;

- Painful, stiff joints
- Small, painless lumps appear over joints
- Tiredness and weight loss
- Fever

In rare cases the inflammation can occur in the lungs or heart this can be serious.

Treatment

Anti-rheumatic drugs are available are tablets or injections. NSAIDS and simple analgesia can be used. Short course of steroids will reduce inflammation although long term use can have serious side effects.

Osteoarthritis

Osteoarthritis is the most common type of arthritis, it causes inflammation of tissues around the joints. It causes damage to cartilage and bony growths appear on the edge of the joints. It mostly occurs in the knees and hips although almost any joint can be affected. It mostly affect women over 50, but also affects men and can occur at any age

Symptoms include:

- Pain, stiffness and difficulty moving joints
- Tender joints
- Joints slightly larger or more 'knobbly'
- Grating or crackling in joints
- limited range of movement
- Weakness and muscle wasting

Treatment

The pain from osteoarthritis is controlled by simple analgesia tablets such as paracetamol, Codeine or NSAIDs. NSAIDs cream is also beneficial as is Capsaicin cream which blocks nervous pain messages.

Gout

Gout is a common type of arthritis affecting mainly men between 40 and 60, usually affects the joint of the big toe but can affect any joint. The affect joint swells and is painful, its caused by a build-up of uric acid in blood that form crystals. Eating too much red meat, seafood and drinking beer worsens the condition as well as being overweight. Diabetes and Hypertension also have an effect.

Treatment

Rest, cool and raise the affect limb this will help reduce swelling.

Non-steroidal anti-inflammatory drugs (NSAIDs) are recommended for treating gout as they relieve pain and reduce inflammation. Corticosteroids can also be used if NSAIDs are not effective

To prevent recurring attacks Allopurinol is usually prescribed which lowers uric acid levels but takes several months to take effect. Drinking plenty of fluids also helps to reduce uric acid levels.

Chapter 8 Allergic Reaction

A mild allergic reaction often causes no more symptoms than an itchy rash, this is not serious in itself, and may be treated with oral or topical antihistamines. If difficulty in breathing or swallowing develops, and/or a sudden weakness regard these as serious symptoms requiring immediate treatment see below.

Anaphylaxis

Anaphylaxis requires prompt treatment its key features which differentiate it from an allergic reaction are;

- Airway swelling
- Airway constriction
- Low Blood Pressure
- Altered level of consciousness
- A sudden onset
- Wheeze and/or Stridor

It can be caused by Insect stings, particularly wasp and bee stings. Certain foods such as eggs, fish, cow's milk protein, peanuts, nuts and drugs including blood products, vaccines, antibiotics and aspirin.

If using any medication it is a wise precaution to prepare for Anaphylactic reactions.

Treatment

- Maintain Airway
- Maintain BP (lay flat, raising the feet

- Administration adrenaline
- Give antihistamine
- Give Nebulised Salbutamol & Oxygen

If using a vial of Adrenaline give 0.5mg intramuscularly(IM) or 0.3mg from an auto injector. The dose is repeated if necessary at 5-minute intervals.

An antihistamine e.g. Piriton can be given orally or by slow intravenous injection of 10mg

An intravenous corticosteroid e.g. Hydrocortisone in a dose of 200 mg can also be given but takes around 6 hours to take effect.

Age Dose Volume of adrenaline
(1 in 1000 (1 mg/mL))

Under 6 months 50 micrograms 0.05 ml

6 months–6 years 120 micrograms 0.12 ml

6–12 years 250 micrograms 0.25 ml

Adult and adolescent 500 micrograms 0.5 ml

These doses may be repeated several times if necessary at 5-minute intervals according to blood pressure, pulse and respiratory function.

An allergic reaction occurs when the body's immune system reacts inappropriately in to exposure to a substance. Histamine is released from tissues acting on blood vessels to causing them to swell (dilate) this is more noticeable in the mouth and face (angio-odema) and blood pressure falls causing patient to collapse.

Adrenaline acts quickly to constrict blood vessels, relax smooth muscles in the airways and improves breathing, it also stimulate the pulse.

Instructions for the administration of adrenaline via the EpiPen® Injector

Grasp EpiPen® in dominant hand, with thumb closest to grey safety cap

With the other hand, pull off grey safety cap.

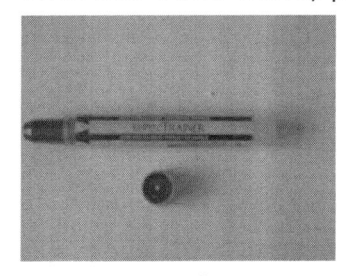

Hold EpiPen® approximately 10cm away from thigh.

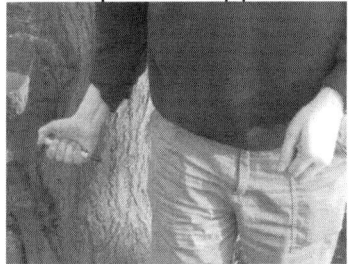

Black tip should point towards outer thigh

Jab firmly into outer thigh so that EpiPen® is at a right angle to outer thigh, through clothing if necessary

Hold in place for 10 seconds.

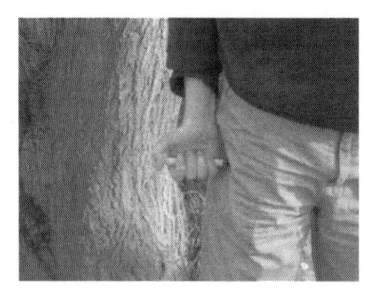

EpiPen® should be removed and handed to team taking over management of the patient.

Massage injection area for 10 seconds

Patient must go to A&E because relapse can occur within a few hours and/or further management may be required

Don't

- Use EpiPen® to practise emergency administration
- Remove grey safety cap until ready to use EpiPen®
- Place fingers over the black tip
- Attempt to inject into vein or buttocks
- Inject into extremities, as adrenaline causes local vasoconstriction

Chapter 9 Respiratory Illnesses

Asthma

Asthma is a disease that affects the lower airways. It is usually triggered by an allergen, exercise or cold weather. It manifests as an obstruction of the airways this is caused by a combination of spasm of the airways causing them to contract and an increase in mucous production that blocks the narrowed airways. It is a common disease affecting over 10% of the population.

Signs & Symptoms include shortness of breath (dyspnoea), increased respiration rate, productive cough and an audible wheeze.

If the attack is serious the patient will not be able to speak in full sentences. Pulse>110bpm and Respiratory rate >25/min

In life threatening asthma, no air movement can be heard when listening to the chest, the patient's extremities become cyanosed (Blue-tinged), there pulse rate drops (Bradycardia) and they become exhausted as breathing becomes more difficult.

Peak Flow Measurement

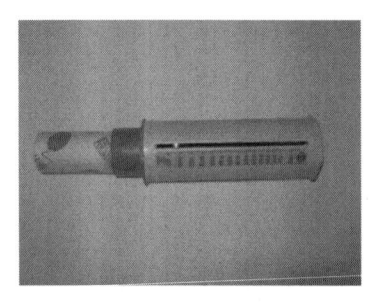

1/Zero the pointer on the scale
2/Breathe in as fully as possible.
3/Blow out as hard and fast as possible.
4/Repeat this sequence twice more.
5/The highest of the three readings is your peak flow.

A person having a serious asthma attack will struggle to blow into the Peak flow meter. Although it is a good baseline measurement for treatment be aware of this.

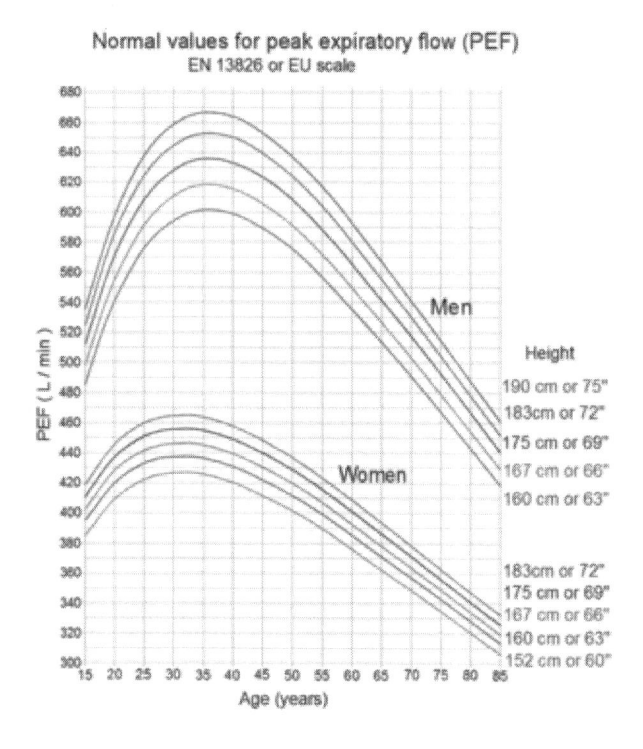

Nunn A, Gregg I (1989). "New regression equations for predicting peak expiratory flow in adults". *BMJ* 298 (6680): 1068-70.

Normal	80-100 % of Predicted Score
Mild Attack	51-79 % of Predicted Score
Severe Attack	33-50 % Predicted Score
Life Threatening	<33 % of Predicted Score

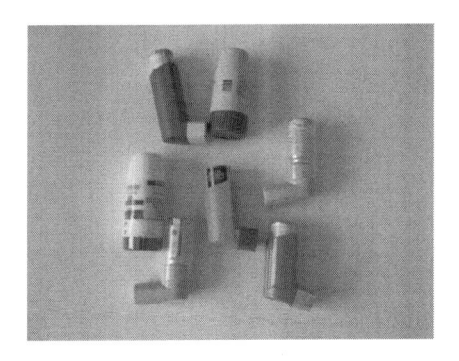

Asthmatics generally use one or more inhalers and possibly nebulised drugs to manage their condition. Inhalers come in all Colours, shapes and sizes, containing different drugs. The drugs fall into two sorts 'preventers' for keeping airways open usually taken at set periods throughout the day and 'relievers' to open the airways when the chest becomes tight or the patient has an asthma attack. The most common inhaler in use is the blue Ventolin inhaler containing the drug Salbutamol used to relieve attacks each puff delivers 0.4mg.

During an attack the patient should be encouraged to use their own inhalers. If they are not gaining relief from them then a nebuliser may help.

Technique for Nebulising

It can be administered via a machine on air or via oxygen from a cylinder. If used on oxygen set the cylinder to deliver 6-8 litres using the dial on top. This is enough to nebulise the medication.

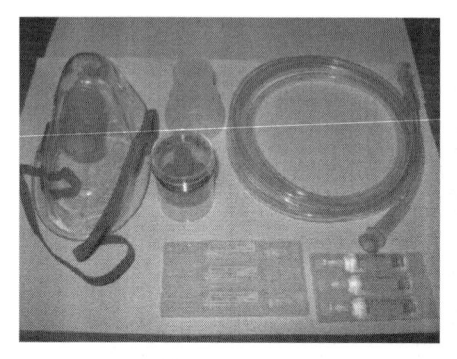

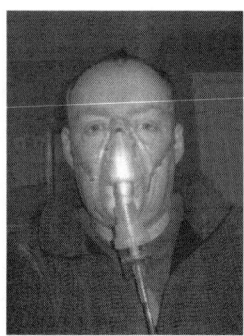

The nebuliser kit consists of a facemask, a two part chamber where the drug is held and oxygen tubing. The drug is placed in the lower chamber and the two halves are screwed together. The mask fits in the top and the tubing in the bottom. Ensure mask fits tightly to patients face. Drugs for nebuliser are usually supplied in plastic ampoules.

The two common drugs used are;

Salbutamol supplied in (2.5mg /5mg) Ampoules

Atrovent supplied in (400mcg) Ampoules

In severe asthma the;

- First dose is 5mg Salbutamol
- Second dose is 5mg Salbutamol + 400mcg Atrovent
- Third + doses is 5mg Salbutamol

If the patient fails to improve and you consider the condition life threatening give;

0.5mg Adrenaline IM

200mg Hydrocortisone IV

40mg Prednisolone by mouth

Following a serious attack if the patient remains symptomatic give 30mg Prednisone for 5 days.

Short courses of steroids have little effect but long or repeated courses have detrimental effect on the body and should be avoided if possible.

COPD

Chronic obstructive airway disease (COPD covers a group of conditions including asthma (see above), emphysema and bronchitis.

Emergency Treatment is the same as for asthma.

Croup

Mild croup requires no specific treatment and can be managed in the community. More severe croup (or mild croup that might cause complications) calls for hospital admission; a single dose of a Corticosteroid (e.g. Dexamethasone 150 micrograms/kg by mouth) may be administered before transfer to hospital. In hospital, Dexamethasone 150 micrograms/kg (by mouth or by injection) or Budesonide 2 mg (by nebulisation) will often reduce symptoms; the dose may need to be repeated after 12 hours if necessary. For severe croup not effectively controlled with corticosteroid treatment, nebulised adrenaline solution 1 in 1000 (1 mg/ml) may be given with close clinical monitoring in a dose of 400 micrograms/kg (max. 5

mg) repeated after 30 minutes if necessary; the effects of nebulised adrenaline last 2-3 hours and the child needs to be monitored carefully for recurrence of the obstruction.

Pneumonia

Pneumonia can be caused by a variety of bacteria and also by aspiration where water or vomit enters the lungs. Treatment is with antibiotics, oxygen if saturations <92% and paracetamol 1g ever 6hrs. Antibiotic therapy depends on strain of bacteria which will not usually be known in the wilderness setting.

Symptoms

Fever, shaking, malaise, anorexia, dyspnoea, cough, thick sputum

Typical dosages for antibiotics would be;

Mild - by mouth

Amoxicillin 500mg - 1g /8 hrs and

Erythromycin 500mg / 6 hr

Mild - Iv

Ampicillin 500mg / 6 hrs and

Erythromycin 500mg / 6 hrs

Severe - Iv

Co-Amoxiclav or (Cefuroxime 1.5g / 8 hrs +. Erythromycin 1g/6hts)

Atypical Pneumonia

Clarithromycin 500mcg/ 12h

Aspiration Pneumonia - iv
Cefuroxime 1.5g/8h & Metronidazole 500mg/ 8h

Avian / Bird Flu (H5N1)

Bird Flu (H5N1) is highly infectious amongst birds but is difficult for humans to catch or spread. It affects wild and domesticated birds. Since 2003 the World Health Organization (WHO) has confirmed over 400 cases of H5N1 in humans, there have been over 250 deaths.

These symptoms are similar to Swine Flu below and can come on suddenly but incubation is usually between 3 and 7 days and last for around a week. Those suspected of having Avian flu should be isolated for 7 to 10 days.

In serious cases, there can be rapid deterioration, with pneumonia and multiple organ failure, and this is generally fatal. For Treatment see Swine Flu below

Swine flu (H1N1)

Swine flu was first identified in Mexico in April 2009, it spread around the world and became a pandemic. By the spring of 2009 the number of cases declined. In August 2010 the World Health Organisation (WHO) declared the pandemic over, although new cases are still being reported in 2011.

Most cases were mild but some deaths occurred amongst patients who already had conditions such as diabetes and that affected their immune or respiratory systems. The "at risk" group includes those over 65, pregnant patients and those with chronic health problems.

A vaccine is available which is currently included with the seasonal flu jab.

Symptoms of Avian & Swine flu include the following, however most are common too season flu.

- Lethargy
- Fever 38C
- Aching Muscles
- Headache
- Runny nose & Sore throat
- Shortness of breath or cough
- Decreased appetite
- Diarrhoea and/or Vomiting

Most patients recover within a week, recommended treatment is to use simple medicines such as paracetamol and aspirin, rest and drink plenty of fluids.

A possible complication is pneumonia, antibiotics are used to combat this.

Anti Virals are rarely needed they will not cure a patient but may;

- Reduce Illness by a day or so
- Relieve some symptoms
- Reduce the chance of serious complications

Two recommended anti-virals are Oselamivir (Tamiflu) and Zanamivir (Relenza)

Acute respiratory distress syndrome (ARDS)

ARDs occurs when the lungs become seriously damaged through severe infection or injury. Cause include serious infection, such as pneumonia or blood poisoning (sepsis). Can also be caused by injuries such as near drowning, smoke inhalation, or severe chest trauma. Even with hospital treatment between a third and a half of patient die.

Symptoms include:

- Cyanosis (blue-coloured lips, fingers and toes)
- Difficult, rapid, shallow breathing
- Increased heart rate
- Lethargy followed by confusion.

Patients with severe cases are generally sedated and ventilated on an Intensive Care unit, less severe may be treated on a ward, both require constant oxygen therapy and treatment of the underlying condition.

Severe acute respiratory syndrome (SARS)

The infection is caused by the SARS coronavirus (SARS CoV) is very serious and affects the lungs. Incubation is between 2 and 7 days.

Symptoms include;

- Temperature of 38°C +
- Lethargy
- headaches

- Fever and chills
- Muscle pain
- Loss of appetite
- Diarrhoea

Between 3 and 7 days after the onset of initial symptoms, it will affect breathing causing; a dry cough, difficulty breathing and a decrease in oxygen levels in the blood which if untreated is usually fatal.

SARS is spread the same way as the flu virus through airborne infections and touching surfaces contaminated with droplets. Being a virus there is no cure for SARS most mentions being ventilated as in ARDS above.

Whooping Cough (Pertussis)

Whooping cough is an infection of the lining of the windpipe (trachea) and the bronchi leading to the lungs. Its highly infectious through droplets on surfaces and in the air from coughing and sneezing. Incidence of cases have been reduced significantly since the introduction of the childhood vaccine. It mainly affect infants and young children but can occur at any age. Incubation can be long 6-20 days before first symptoms appear.

- Mild Pyrexia
- Runny nose and sneezing
- Watering eyes

- Dry, irritating cough and sore throat

After 1 or 2 weeks more severe symptoms appear

- Main symptom is hacking cough with 'whoop' sound
- Cough is productive with thick phlegm
- Fatigue from strain of coughing
- Vomiting

Coughing comes in bouts roughly 12 to 16 per day

Complete recovery can take up to 3 months with frequency of bouts decreasing over that time.

Treatment

If diagnosed early within first two weeks it can be treated with antibiotics but they have little effect once the infection is established. Also rest and drink plenty of fluids.

Chapter 10 Abdominal illnesses

Appendicitis

The appendix is believed to be the remains of an additional piece of intestine, which through evolution became redundant. It has also been suggested it has a role in the immune system. But no disability has been noted in the routine surgical removal of it.

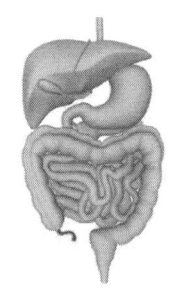

© 3drenderings | Dreamstime.com

The appendix is a tube approximately 10cm long with the distal end closed. It is connected to the caecum and is located near the junction of the small intestine and large intestine.

Appendicitis is inflammation of the appendix. It causes pain, nausea, vomiting, anorexia and a fever. Constipation is usual but diarrhoea may also occur. A fast pulse is an indication of any infection in the body.

It is estimated that 10% of the population will have Appendicitis at some point, the peak being in males between 10 and 20 years of age.

A life threatening problem will occur if the appendix ruptures and faeces enters the peritoneum - this will cause peritonitis (an inflammation of the peritoneum).

The pain from appendicitis is characterised by tenderness and rigidity (guarding) and rebound (press then let go) tenderness in the right iliac fossa.

Laughing, coughing or movement may aggravate the pain.

Unfortunately an abdominal exam may not always show the signs mentioned above particularly if the caecum is distended or the appendix lies within the pelvis.

Treatment

The usual treatment is the surgical removal of the appendix (Appendicectomy). This is not generally an option in the field, however some have been done as the importance of early surgery is high. In remote locations antibiotic, analgesia and symptom control are the best option given limited resources.

All medication should be IV or PR as oral medication might not be absorbed effectively if the problem you are treating is abdominal. Recommended antibiotics are Cephalosporin and Metronidazole 1g twice per day

Cholecystitis

Cholecystitis is inflammation of the gallbladder, which is a pouch that lies under the liver on the right side of the upper abdomen (RUQ) it stores bile which is produced in the liver and used in the gut to aid digestion. The cause of Cholecystitis is often attributed to gallstones but can be unknown. It can affect anyone but is more common in women.

Treatment

First line treatment for Cholecystitis is to stop the patient eating and drinking, if the inflammation is caused by gallstones there will often fall back into the gall bladder and the inflammation will subside. The patient will need IV fluids and painkillers during this period. If an infection is suspected antibiotics can be given this way as well.

If the condition is severe or repeated the gallbladder is removed surgically, this has little detrimental effect other than some discomfort when eating very fatty meals, which the stored bile would have helped digest.

In severe cases the gallbladder becomes severely infected and sometimes gangrenous this can lead to blood poisoning (septicaemia).

Diverticulitis

Diverticulitis is a condition where small pouches form and protrude from the colon through weaknesses in the intestinal wall, If these diverticula become inflamed this is called diverticulitis.

It's a very common condition and generally affects people over 40, it is estimated that 50% of the population will have diverticula by the time they are 50 and 70% before the age of 80. Although many people remain asymptomatic.

Symptoms can include;

- Severe Pain below navel, moving to the lower left quadrant of the abdomen.
- Pyrexia
- Frequent and/or painful urination
- Nausea and/or Vomiting

- Constipation
- Rectal Bleeding

Treatment

A high fibre diet will help prevent and treat the condition. Symptomatic medication may be needed to control, pain, Nausea, Vomiting, Constipation etc. Acute diverticulitis is treated with antibiotics to cure infection and inflammation.

Complications

In severe cases rectal bleeding may be severe and patient will need transfusion.

Abscess may form. Colon can be narrowed causing blockages, Fistulas can form between parts of digestive systems. Perforation of the colon often leads to peritonitis and sepsis.

Irritable Bowel Syndrome (IBS)

Irritable Bowel Syndrome affects 20% of the population at some point in their life, the cause is not fully understood but bouts are believed to be triggered by a number of causes including infections, nerves in the gut and intolerance to foods. But it often occurs in people with a normal digestive system which is free of disease and infection.

Common Symptoms include;

- Abdominal pain
- Bloating
- Diarrhoea and/or constipation
- Pellet or ribbon like fasces
- Mucous in faeces

Rare symptoms may include;

Nausea, headache, belching, loss of appetite, tiredness, backache, muscle pains, feeling quickly full after eating, indigestion and frequent and/or painful urination.

Treatment

There is no cure and bouts may vary in length, symptoms and severity. Treat symptom if severe as they occur.

The use of a high fibre diet is controversial as may irritate the bowel more. Try and increase soluble rather than insoluble fibre. Sources include oats, ispaghula (psyllium), nuts and seeds, some fruit and vegetables and pectins. Avoid coin and wheat bran.

Peritonitis

Peritonitis is inflammation of the peritoneum which is a membrane sac covering the abdominal organs due to a infection.

The peritoneum is usually free of germ-free so doesn't have capacity to beat of infection which can spread to the circulation system causing blood poisoning (sepsis) then to other organs causing organ failure and death.

Peritonitis can occur directly in the peritoneum or via damage to another organ allowing faeces to contact the peritoneum such as in a ruptured appendix or tears to the colon in diverticulitis.

Examine the abdomen it should move with breathing, ask the patient to 'suck' stomach in on inspiration then 'blow' it out on expiration. If the patient has peritonitis they will splint the abdomen and it will not move with breathing. In fact they will be trying to lie as still as possible as any movement will cause severe pain.

Peritonitis is very serious and 10% of patient die even with full medical support in hospital. It's also very painful, patients will require analgesia and antibiotics or antifungal to fight the infection. Surgery is sometimes required to remove infected tissue.

Urinary tract infection (UTI)

Urinary tract infection are very common in women, but very rare in men. The severity and treatment depends on where in the urinary tract they occur.

Lower UTIs either affect the bladder and is called cystitis, or the urethra which known as urethritis. Upper UTIs either affect the kidneys or ureters and are the more serious of the two types.

Rare complications of untreated UTIs include Kidney failure and sepsis.

Symptoms of a Lower UTI include,

- Burning pain during urination
- Increased frequent
- Lower abdominal pain
- Cloudy, Smelly Urine

- Back pain or Tenderness in the pelvic area

Symptoms of a Upper UTI include,

- As above plus
- Temperature 38°C +
- Fever
- Nausea and/or vomiting
- Diarrhoea
- Pain may be in your side, back or groin.

Treatment

Treatment of Lower UTI is with a 3 to 7 day course of trimethoprim

Treatment of an Upper UTI is with 7 to 14 day course of Co-Amoxiclav or Ciprofloxacin

In either case take analgesia for pain, take plenty of fluids, cranberry juice is believed to be beneficial.

Kidney Stones

The kidneys process waste from the body and it is eliminated via urine. Sometimes the waste can form into crystals, these join together to form hard kidney stones. There are fairly common affecting more men than women at a ratio of 4:1 usually between the ages of 30 and 60. The pain from kidney stones is called renal colic. Stones are passed from the body in urine, but larger ones can become lodged in the urinary system. This is very painful, can lead to infection and damage of the kidneys. The patient can suffer all the symptoms listed under UTIs and blood may be present in the urine if the tract is scratched.

Treatment

Very small stones are usually passed without problem, medium ones may still be passed but will cause pain as they travel through the tract, analgesia may be required as well as medication for nausea Larger stones are usually broken down by ultrasound or removed by surgery. Some stones consisting of uric acid these are softer than other types and can be dissolved by drinking 3 or 4 litres of fluid a day and taking medication to make urine more alkaline.

Enlargement of Prostate

The prostate is a small gland located in the pelvis, between the penis and bladder, and surrounds the urethra. If it becomes enlarged it can place pressure on the bladder and urethra.

Symptoms include;

- Difficulty urinating
- Increased frequent of urination
- Difficulty emptying the bladder

Treatment

Avoiding alcohol and caffeine, exercise regularly. Medication options are Finasteride which shrinks prostate enlargement. Tamsulosin is an alpha blocker that is widely used to treat prostate enlargement by relaxing bladder muscles.

Chapter 11 Neurological Problems

Seizures

Can vary in intensity from brief periods of absence where the patient appears to be distracted to prolonged full body convulsions and repeated fits. Many things can cause seizures; epilepsy, CVAs, Head and Spinal Injuries as well as other medical problems.

Three types of Seizures exist;

- Absences (Petit Mal)
- Tonic Clonic (Grand Mal)
- Focal (Jacksonian)

Focal seizures may only affect a single limb and do not necessarily require emergency treatment.

If the patient is known to have seizures and they are generally short in duration (<1 min), let the seizure run its course, protecting the patients head, don't try and restrain them and never put anything in their mouth.

A Tonic Clonic Seizure generally follows four stages;

- Aura (Indication of Seizure)
- Tonic (Patient becomes Rigid)
- Clonic (Seizure activity)
- Post Ictal (Recovery Phase)

If the fit is prolonged treatment should be give;

- Administer Oxygen if required
- Administer Diazepam (Rectal or IV)

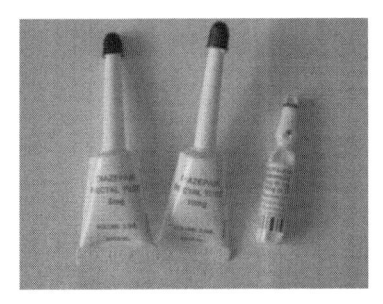

If the seizure is caused by a high temperature "a febrile convulsion" try and cool the patient by;

Removing clothing

Cool Sponging

Paracetamol (Tablet, Liquid, IV or Suppositories)

If a patient remains in a state of seizure or has multiple seizures for a prolonged period they are at serious risk of sustaining a long term brain injury. Fits lasting over an hour have an 80% chance.

Always check the casualty's blood Glucose measurement and supplement if required, as seizure activity burns lots of energy.

If Trismus (Clenched teeth) compromise patient airway, consider using a Nasopharyngeal Airway (NPA).

Stroke (CVA)/TIA

A stroke or Cerebral Vascular Accident (CVA) is an interference with normal brain function caused either by a clot (85%) or bleeding (15%). Depending on the severity and location of the damage the patient will exhibit different signs and symptoms. They may lose the use of their arms and legs on one side of the body, or there may be a decrease in strength and power on the affected side or just a heavy feeling in the limbs. Patients can become disorientated and confused, speech may be slurred, absent or they become dysphasic losing the ability to select appropriate words to form sentences. They may have unequal pupils or develop a facial droop. A headache and high blood pressure often are present. Treat the patient symptomatically if oxygen saturation drop below 95% give oxygen if available. They may also need analgesia and antiemetic medication.

A Transient Ischemic attack is a small stroke where the symptoms resolve without intervention normally within 24 hours and frequently within minutes or a few hours. Although separate parts of function may return at different times.

A simple way of assessing a patient you think may have had a stroke is through the FAST acronym. It stands for FACE, ARMS, SPEECH, TEST.

FACE

Look for facial droop, dribbling etc

ARMS

Ask them to squeeze both your hands with theirs, feel for a marked difference in strength between body sides. Ask if they can raise their legs and assess in the same way.

SPEECH

Listen for slurring, confusion, using inappropriate words. If administering first aid, the patient should be supported in a comfortable position, where any dribbling can drain away. Administer oxygen in saturations below 95%.

The definitive treatment for CVAs caused by clots is to use a thrombolytic (Clot Busting) drug. This is only usually given after the patient has had a CT Scan to eliminate a bleed as the cause as giving thrombolytic to someone who has a bleeding in the brain will worsen their condition.

Before a patient is eligible for a CT Scan and therefore potentially for thrombolytic therapy they need to meet certain criteria. This varies between hospitals but as a general guideline the following apply;

- Patient aged between 18-80
- Not taking Warfarin (A blood thinner)
- Not had a seizure
- Showing signs of a stroke

There have also been some trials using GTN spray to lower blood pressure of patients with suspected strokes. Although effective in strokes caused by clots, it will worsen strokes caused by bleeding.

Without the benefit of CT Scans, a decision would need to be made to weigh the risks versus the potential benefits of using either drug.

Following a CVA patients are often prescribed Warfarin or Aspirin to help keep their blood from clotting. Patients taking Warfarin need the clotting factors in their blood checked regularly to ensure dosage is correct. This is not appropriate if in a remote location.

Migraine

This is a severe throbbing headache usually felt across the temple or at the side of the head. Nausea or photophobia may also be experienced. Migraines are common affecting about 1 in 6 of the population mostly women. Around a third of sufferers experience an aura, which is a warning sign that an attack is imminent, auras vary between people. Frequency of attacks can vary between several attack a week to once every few years.

Treatment

Simple analgesia is often all that is required, if nausea is a problem and the patient cannot swallow pills, take bucastem this is an anti-nausea medication that dissolves in the mouth. Another option is to take analgesia in suppository form. Tablets are also available that contain both ant-nausea medication and analgesia.

Another option is to use sumatriptan this causes blood vessels around the brain to contract this is the opposite effect to that the dilation caused by the migraine. Available as tablets, injections and nasal sprays.

Chapter 12 Circulatory Problems

Heart Attack (Myocardial infarction) (MI)

A heart attack occurs when one of the coronary arteries supplying the heart with blood gets blocked. The effect is dependent on where the blockage occurs, if it occurs in one of the smaller side branches, the part of the heart muscle (myocardium) becomes starved of oxygenated blood and dies. If only a small non vital area is affected the patient may recover without any treatment and only minimal effect. If however the blockage occurs at the top of one of the main arteries, it's likely the conduction system of the heart will be thrown out of sync and the patient will immediately go into Cardiac Arrest. Most heart attacks occur between these two extremes. See below for a diagram of the heart showing the system of coronary circulation and plaque formation where a build up of cholesterol occludes the Vessel.

Statistically only 1% of people presenting with chest pain symptoms to the emergency services are actually having a heart attack. They typically present with a central heavy, crushing chest pain, which may radiate to their left arm and/or neck. I say typically as around 30% of people describe the pain as sharp rather than crushing and a number of other have an epigastric pain which is often mistaken for indigestion and vice-versa. In a small percentage of cases the pain may be in the neck or arms only and can be described as a sore throat or heavy arm. Some patients will not experience any pain, this is more common amongst those that have diabetes due to degradation of nerves.

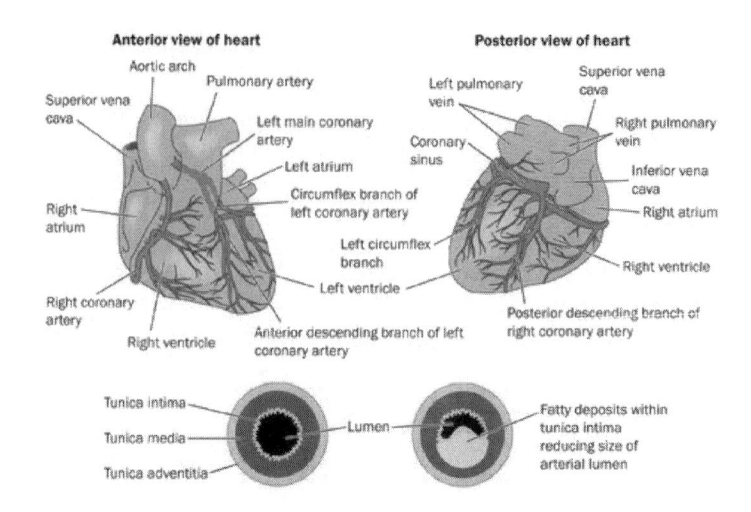

Depending on the intensity of pain and the nature of the damage to the heart the pulse and Blood Pressure may be raised or lowered and the pulse may become irregular. Oxygen saturations may also be decreased if the heart is struggling to pump blood around the body.

Other signs and symptoms can include nausea, vomiting, palpitations, shortness of breath and a feeling of dread. As you can see there is no such thing as a typical heart attack!

Two more definitive tests are an ECG recording and a blood test to measure changes in cardiac enzymes in the blood. Even these although normally reliable are not infallible.

Therefore the diagnoses of a MI is based on all the above and the history of the patient and onset of the symptoms.

The management can be broken down into sections dependant on the situation, knowledge and resources available.

From a first aid perspective where emergency services are available once a MI is suspected the patient should be reassured, sat down and given 300mg of aspirin providing they are not allergic to it. Try and keep the patient sitting up to assist with their breathing, but if they are very pale try to raise their legs to improve circulation. If they become unconscious place them in the recovery position.

Talk to the patient and try and get as much information about what happened, their medical history and details of any drugs they take and allergies they have. This will help the Paramedic particularly if the patient becomes unconscious before they arrive.

If help is delayed or not available the next stage involves measures to reduce symptoms and strain on the struggling heart.

If you can measure the patient's oxygen saturation give supplementary oxygen to maintain a saturation level of 94% or above. If the patient also has COPD then saturations can be maintained in the 88%-92%

range. If saturations are above 94% oxygen is not required. If oxygen saturation monitor is not available give 100% oxygen.

For more information on Oxygen therapy See the British Thoracic Society website.

The second drug used is Glyceryl Trinitrate (GTN)

The third action is to use a fast acting analgesia the two commonly used are;

Entonox (also called Gas and Air / Laughing Gas) or preferably an Intravenous pain killer such as Morphine.

The use of an anti-emetic (anti-sickness) drug may also used if the patient is nauseous. The administration of morphine may also induce nausea.

Suitable anti-emetic drugs are Metoclopramide, cyclizine or Prochlorperazine.

An acronym for remembering the treatment for a Heart Attack is MONA

(M)orphine, (O)xygen, (N)itrates and (A)spirin

Pre-Hospital / Hospital Care

There are two main routes of definitive treatment for an MI

The first is a Stent, which is a metallic mesh tube used to re-open a Coronary Artery allowing blood to move freely to the affect part of the heart. This would require a hospital so not applicable to immediate care in a survival situation.

The second is Thrombolysis which is the use of a clot busting drug such as Reteplase or Tenecteplase. Drugs of this type have some serious side effects and contraindications and are also very expensive £600-£800 a dose.

It is also possible to thrombolyse a patient then insert a stent.

Additional Therapeutic drugs

In addition to the standard treatment detailed above the Wilderness Medicine Society recommends the following drug therapy for patients post Myocardial infarction.

Immediately give 300mg Clopidogrel in addition to 300mg Aspirin.

Give 0.4 mg GTN by tablet or spray every 10 minutes as long as Systolic Blood Pressure is above 100mmHg.

Give 25mg Metaprolol or 25mg Atenolol four times a day starting 6 hours after the pain started. Unless the Systolic Blood Pressure is below 100mmHg or pulse <60 bpm.

Angina

Angina is a disease where the coronary arteries have become narrowed due to the build up of cholesterol in the arteries supplying the heart with blood. The patient experiences pain as the myocardium is starved of oxygen rich blood. There are two types of angina, stable and unstable.

In stable angina the patient generally knows what amount of exercise they can comfortably do without over exerting their heart. They are able to live their life within these boundaries. If they exceed their normal exercise tolerance they experience pain. This is relieved by using GTN spray or tablet which allows more blood to get to the heart muscle.

Deep Vein Thrombosis (DVT)

Deep vein thrombosis (DVT)are blood clots in deep veins particularly in the calf and the thigh, they affect around 0.1% of the population each year and are more prevalent in obese people and smokers. The main complication is that part of the clot may detach and travel to the lungs causing a pulmonary embolism.

Symptoms include;

- Heavy aching Pain

- Swelling and or Redness

- Warm skin

- Tenderness particularly on the back of leg below knee

Treatment

Pulmonary embolism are treated with anticoagulant usually injected Heparin and sometimes oral warfarin.

Heparin comes in two types standard and low molecular weight heparin (LMWH). Both are given intravenously. LMWH is more reliable and requires less monitoring and has less serious side effects it can also be given as subcutaneous injection (Into the skin)

Warfarin may be needed to be taken for a long period after the heparin injection sometimes indefinitely to prevent further blood clots occurring. Anybody taking warfarin will need the clotting factors in their blood checked every four weeks to ensure its at the correct level. Patients on warfarin are at a greater danger of blood loss as it doesn't clot as quickly. For pregnant women, heparin injections are given over the treatment period, instead of warfarin.

In a remote situation where warfarin levels cannot be measured take 75mg aspirin and/or clopidogrel instead.

Compression stockings help prevent complications and should be worn until you go to bed every day for two years follow A DVT, they can be removed if you keep your legs raised whilst resting, but ensure your foot is higher than your hip. Raising the foot of your bed also helps

Pulmonary Embolism (PE)

A pulmonary embolism is caused by a blood clot lodging in a blood vessel within the lung like a heart attack affect depends on where the clot occurs, however any pulmonary embolism is a serious condition.

Risk factors for pulmonary embolism include a history of DVTs, use of contraception or long haul flights. Around 0.1% of the population develop DVTs and 10% of people with untreated DVTs develop pulmonary embolisms

Most cases of pulmonary embolism develop when part, or all, of the blood clot travels in the bloodstream from the deep veins in your leg and up into your lungs.

Symptoms are;

- Breathlessness that can come on gradually or suddenly
- Sharp Pain which can be pin pointed
- Pain can increase on Inspiration

Treatment

Pulmonary embolism are treated with anticoagulant usually injected Heparin and sometimes oral warfarin.

Heparin comes in two types standard and low molecular weight heparin (LMWH). Both are given intravenously. LMWH is more reliable and requires less monitoring and has less serious side effects.

Warfarin may be needed to be taken for a long period after the heparin injection sometimes indefinitely to prevent further blood clots occurring. Anybody taking warfarin will need the clotting factors in their blood checked every four weeks to ensure its at the correct level. Patients on warfarin are at a greater danger of blood loss as it doesn't clot as quickly. For pregnant women, heparin injections are given over the treatment period, instead of warfarin.

In a remote situation where warfarin levels cannot be measured take 75mg aspirin and/or clopidogrel instead.

Chapter 13 Diabetes

There are currently over 2.3 million people with diabetes in the UK and there may be up to another 750,000 people who have the condition and do not know it. It has now become a standard test if you have contact with medical professionals.

There are two different kinds of diabetes mellitus these are commonly known as type 1 or type 2 diabetes. The type you have determines the body's ability to produce and metabolise the hormone insulin. Insulin is required for the body to make use of sugar in our diet. Most bodily functions rely on fuel in the form of fat and sugar to work. The brain however relies on sugar only.

Type 1
Type 1 diabetes affects between 5 – 15% of people that have the disease. Type 1 diabetes was formerly called early onset diabetes as it was developed in childhood or before the age of 40. These patents are unable to produce their own insulin or there is a problem metabolising sugar in the body.

This type of diabetes is therefore controlled with sub-cutaneous insulin injection and regular blood glucose monitoring. The patent learns to vary the amount of insulin they need based on what they eat, their level of activity and blood glucose measurement.

Type 2

Type 2 diabetes usually occurs later in life, when the body can still make some insulin, but not enough, or when the insulin that is produced does not work properly (known as insulin resistance). Triggering factors include obesity, lack of exercise, poor diet, high blood pressure, high cholesterol and genetic disposition. It can develop earlier in life if these risk factors are prevalent. The onset of Type 2 diabetes can be delayed with lifestyle choices, but is incurable once developed and the condition is on the rise. Type 2 diabetes is controlled by diet, exercise, tablets and sometimes insulin.

Treatments

Insulin

Insulin is a hormone which enables cells to absorb glucose from the blood stream which can then be distributed around the body via the circulatory system. Predominantly used to treat type 1, insulin may be prescribed to type 2 patients who are deteriorating or not responding to anti-diabetic meds.

If the body is unable to metabolise the sugar it stays in the blood stream, this can damage the vessels and organs it comes in contact with.

Diabetes Medication

Metformin - this is often the first medicine that is advised for type 2 diabetes. It mainly works by reducing the amount of glucose that your liver releases into the bloodstream.

Gliclazide - increase the amount of insulin produced by your pancreas. They also make your body's cells more sensitive to insulin so that more glucose is taken up from the blood.

Acarbose - this slows down the absorption of carbohydrate from the stomach and digestive tract, preventing a high peak in the blood glucose level after eating a meal.

Nateglinide and Repaglinide stimulate the release of insulin by the pancreas. They are not commonly used but are an option if other medicines do not control your blood glucose levels.

Thiazolidinediones (glitazones) (e.g. pioglitazone, rosiglitazone) - These make the body's cells more sensitive to insulin so that more glucose is taken up from the blood. They are a third line treatment for people who do not respond to other treatments or in whom other treatments are not suitable.
These are the main medications prescribed for type 2.

Glucose monitoring

Only a drop of blood is needed to gain a blood sugar reading. This is usually taken from the end or side of a finger. The normal blood glucose level is between 4 - 7 mmol/l before meals, and less than 10 mmol/l two hours after meals, mmol/l means millimoles per litre and is a way of defining the concentration of glucose in the blood.

Diet/lifestyle modification.
Diabetics need to modify their diets to reduce the

amount of raw sugar intake. Other, more complex sugars (carbohydrates) are broken down more steadily by the body and are much less detrimental. Exercise is also beneficial as this helps the body to burn off excess blood glucose.

Symptoms
In both conditions the symptoms are essentially the same. The blood sugar begins to rise (hyperglycaemia) and results in Lethargy, with aching legs being the first indicator.

The next symptom will be increased urine output coupled with extreme thirst. This is because the glucose left sitting in the blood has to be discharged via the kidneys.

After that the patient will rapidly begin to lose weight as the body will have burnt up all of its fat reserves, and become much more lethargic due to the decrease in reserves.

The next stage comes about because the body begins trying to burn protein in the form of muscle-mass as a last-ditch attempt at fuelling itself and the patient will literally waste away, stones lost in days. The by-products of this process are called ketones. If ketones are in the blood then they turn acidic and rapidly bring about a condition called ketoacidosis,

The patient's breath may start smelling of ketones which is a 'fruity' or 'pear drops' smell similar to

Acetone, however this is a well documented but an unusual symptom.

Emergency Treatment of a patient suffering from Hypoglycaemia 'Hypo'

This is when the blood glucose level falls too low. The actual level at which the patient become symptomatic varies between cases but a reading below 3.5 mmol/l is concerning.

A hypo occurs when blood glucose drops to such a point that there is not enough to fuel the brain. This has various causes.

Too much medication,
- Not enough food eaten,
- Over exertion,
- Illness

Or a combination of any or all of these.

A patient suffering from a 'Hypo' will gradually become more confused, irritable, aggressive, disorientated and uncooperative. A diabetic 'Hypo' is often confused for intoxication. This can rapidly lead to unconsciousness and seizure, followed by coma and then death.

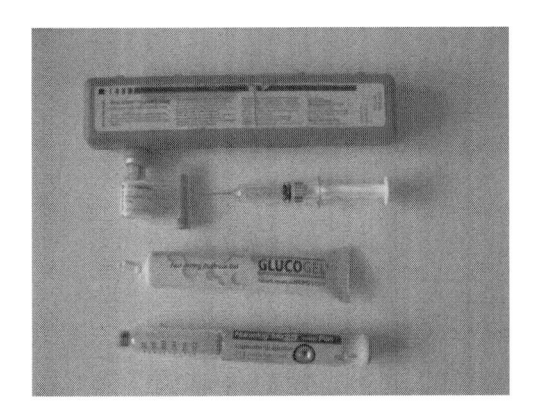

If the casualty is still conscious get them to take sugar in the form of glucose tablets, jam or non-diet drinks. They may have Glucogel (Hypostop) gel which is concentrated sugar.

The information given is for both first aiders and as a suggested treatment for extreme survival situations. In these circumstances gold standard treatment is not always available due to lack of skill, equipment or other resources.

The best practice for the treatment of an unconscious diabetic is to administer bolus doses of intravenous glucose usually 1 or 2 100ml doses. The amount that needs to be given is titrated (Calculated) based on the patient's blood sugar, response and level of consciousness.

In order to administer this you would need all the cannulation equipment, a giving set and the glucose which is a prescription only drug (POM). Additionally you would need to know how to take a blood sugar measurement and have the clinical knowledge on how to interpret and act on the results. Further you would need to know how to cannulate safely as injecting glucose incorrectly can cause gangrene which could lead to loss of a limb.

Therefore a glucagon intramuscular injection is a preferred method for teaching to lay people as it is difficult to get wrong and has relatively few side effects compared to IV glucose. It does not involve IV access so has much less risk of introducing infection. Glucagon is also exempt from prescription only (POM) status if administered in an emergency situation so can be done legally by a non HCP. It is taught to carers of diabetics and community first responders for these reasons.

Glucagon releases emergency glucose stored in the body however if the hypoglycaemia was caused by exertion it is likely the patient has already used their stored glucose and the treatment will not work. In this case IV glucose is the only option available

Chapter 14 Infectious Diseases

Anthrax

Anthrax is a bacterial infection, which is rare and mainly occurs in Africa, Asia, China and Eastern Europe. It has also been used as a bioterrorism weapon. Spread is by contact with the infected carcasses of hoofed animals such as goats and sheep. Three routes of infection;

- Through a cut or puncture in the skin.
- Inhaled spores through the Lungs.
- Through eating poorly cooked infected meat.

Black pustules form with swelling, fever, enlargement of liver and spleen, if inhaled can cause pneumonia and shortness' of breath, if swallowed can cause internal bleeding. Treat with Ciprofloxacin 500mg BD for up to 60 days if Lungs or stomach are involved use ciprofloxacin 400mg IV BD and Clindamycin 900mg IV TDS + Rifampicin 300mg IV BD. Change to oral combination when suitable.

Cholera

Cholera is caused by faecal-oral contamination presents with copious watery stools, fever and vomiting, which leads to profound dehydration. Incubation ranges from a few hours to 5 days. Antibiotics are not usually effective, treat diarrhoea and rehydrate patient.

Malaria

Malaria is spread via mosquito bites and on rare occasions by blood to blood contact. There are different strains of Malaria the most virulent being Falciparum malaria. The disease has a variable incubation period usually from 7 days to one month but can appear much later, so patients often don't become ill until after they have returned from travelling.

Patients experience some or all of the following;

Fever, shivering (Rigors), muscular pain, headache, diarrhoea, nausea, cough, lowered blood pressure, loss of appetite and general fatigue.

In severe cases patient may become jaundiced, lose consciousness, experience seizures, have lowered blood sugar, become shocked and develop breathing problems.

Prophylaxis is certainly better than treatment, if you are unlucky enough to contract it the type can be determined using a text kit from BinaxNOW taking approximately 10 minutes. They also do a range of test kits for a variety of other conditions. The test kit has a shelf life of 12 months but is expensive, with a box of 25 Kits costing around £340.

The Health Prevention Agency (HPA) in the UK publishes guidelines which are available on-line for Prophylaxis showing which areas have a high risk of malaria and where the strains are resistant to the various drugs.

300mg Chloroquine / Week

200mg Proguanil / Day

300mg Chloroquine / Week plus 200mg Proguanil / Day

250mg Mefloquine / Week

100mg Doxycycline / Day

250mg Atovaquone plus100mg Proguanil / Day

Treatment

Non-Falciparum Malaria

Give 600mg Chloroquinine, followed by 300mg at 6, 24, 48 hours. Followed by 30mg/day Primequinine for 14 days.

Falciparum Malaria

Give 600mg Quinine TDS plus 200mg Doxycycline for 7 days.

Measles

In our modern world most childhood diseases that were common years ago have all but disappeared in the UK. Worldwide 800,000 children die from measles each year, the majority of which live in developing countries. It is the leading cause of blindness in African children.

In 2005 there were only 77 cases of measles reported in the UK. Unfortunately a year later this had risen to 449 cases. A boy's death from the disease in April 2006 was the first UK fatality in 14 years. The rise in cases was due to parents who were worried about side effects of vaccines not getting their children vaccinated with the triple MMR (Measles, mumps and Rubella). Measles is probably the most infectious disease in the UK and it is hanging around waiting for enough people without immunity to cause an epidemic. Post fall we won't have access to vaccines which are difficult to produce and have a relatively short lifespan of 18 months and require storage between +2°C and +8°C, costing around £12 but it's a POM so almost impossible to obtain unless you're a doctor.

Measles has been a notifiable disease in England and Wales since 1940, and notifications varied between 160,000 and 800,000 cases per year, the peaks occurring in two-year cycles. Before the introduction of measles vaccination in 1968, around 100 children a year in England and Wales died from the disease.

Adolescents and adults born in 1970 or later who have not been immunized with MMR vaccine, should receive two doses of MMR separated by at least a month. Persons born before 1970 are likely to have had measles as a child and do not normally require vaccination.

Early warning signs last 2 to 4 days and include irritability, a runny nose, conjunctivitis (pink eye), a hacking cough and an increasing fever that comes and goes. These symptoms may last up to 8 days.

Additionally, 'Koplik' spots which are highly characteristic of the early phase of measles may be spotted in the mouth on the inside of the cheek opposite the 1st and 2nd upper molars one or two days before the rash appears. The spots look like tiny grains of white sand, each surrounded by a red ring. However, these are difficult to see.

The typical measles rash appears from day four with the fever usually peaking at this time. The rash usually starts on the forehead and spreads down over the face, neck and body and consists of flat or raised red or brown blotches which flow into each other. The rash lasts for between four to seven days.

Symptoms develop 9 to 11 days after becoming infected and last up to 14 days from the first signs until the end of the rash. However, the illness can last longer if complications develop.

Measles is highly contagious and can be transmitted from four days prior to the onset of the rash to four days after the rash appears. It is most infectious just before the rash appears and so people tend to spread the virus before realising they are infected. It is estimated that 90% of susceptible close contacts of someone infected with measles will also become infected with the virus.

This would mean that a person could be contagious before any symptoms appear at all.

The virus resides in the mucus of the nose and throat of someone infected and is spread by coughing and sneezing which causes droplets of infected mucus to be sprayed into the air. These droplets are then breathed in by others or transferred to the mouth by fingers which have handled a surface contaminated by infected droplets. The virus remains active and contagious on surfaces for up to two hours.

After exposure to the virus it usually takes around 10 days (7-18 days range) for the first symptoms to appear.

In the UK, complications are quite common even in healthy people and approximately 20% of reported measles cases experience one or more complications. Complications are more common among children under 5 years of age, those with weakened immune systems, children with a poor diet and adults.

Common complications of measles include:

Diarrhoea and vomiting, conjunctivitis (eye infection), laryngitis (inflammation of the voicebox), inner ear (otitis media) infection, febrile convulsions, pneumonia.

Less common complications are:

Meningitis, hepatitis (inflammation of the liver), encephalitis (inflammation of the brain) which can lead to permanent deafness, mental retardation and can be fatal, bronchitis and croup, squint as the virus can affect the nerves and muscles of the eye.

On rare occasions measles can lead to:

Serious eye disorders, heart and nervous system problems, and very rarely (1 in 8000 cases in children under two and 1 in 25,000 cases in older persons), a progressive and fatal brain infection called subacute sclerosing panencephalitis (SSPE) sometimes many years after the first bout of measles. Death occurs in 1 in 5000 cases.

Catching measles in pregnancy can also cause miscarriage, premature labour or a baby with a low birth weight.

There is no specific treatment for measles, but drinking lots of clear fluids to prevent dehydration and taking Paracetamol to reduce the fever are recommended. As measles is caused by a virus, antibiotics are ineffective although these may be prescribed if a secondary bacterial infection develops.

Meningitis

Meningitis is inflammation of the meninges which covers the spinal cord and the brain. Signs and symptoms include a sudden onset of fever, severe headache, photophobia, neck stiffness and a rash which doesn't disappear when pressed on with a glass. Transmission is via droplet secretions, incubation period is between 2 and 10 days. Patients are no longer infectious 24hrs after start of therapy. Untreated it can progress to seizures, coma and death. Immediately treat with IV Benzylpenicillin 1.2g or IV Cefotaxime 2g.

Plague

Plague is caused by a bacteria that infects wild rats, it is rare but can be found throughout the world. Incubation period is 2 to 6 days. It comes in two forms, bubonic plague (spread by infected fleas) and pneumonic plague (spread by droplets from coughing). Treat with Doxycycline 100mg BD

Rabies

Rabies is spread through the bite of an infected animal. Any bite in a high risk area from an unprovoked animal should be considered to be at the risk of being rabid.

Incubation is usually 1 to 2 months but can vary between months and years. Symptoms include headache, confusion, fever, hallucinations and hydrophobia (fear of water). Paralysis may also occur.

Treatment starts with aggressive cleaning of the wound using alcohol or providone-iodine followed by a tetanus toxoid injection.

Followed by rabies immune globulin 20 IU/Kg with half infiltrated into wound and the other half by gluteal injection (upper outer quadrant of buttock). Human diploid cell vaccine (HDCV) 1ml IM on days 0,3,7,14,28

Tetanus

Tetanus is caused by the bacteria Clostridium Tetani. It is transmitted through a puncture wound. Onset can take up to two months, but usually takes around 7 days. Early sign is jaw stiffness (hence the common name Lock Jaw), muscle stiffness, headache, fever and spasm. It is fatal in 40% of cases, death occurring due to failure of respiratory muscles. Treatment is with anti-toxin and symptomatic relief, effects may last several weeks. Tetanus immunisations last 10 years and cover for life is received after 4 injections unless placed in a high risk situation.

Tuberculosis (TB)

Tuberculosis (TB) is very infectious and transmitted by droplet infection. A primary pulmonary lesion usually leads to lymph node involvement. There may be no initial symptoms but the patient may be symptomatic from the lung lesion and have a cough, thick sputum, haemoptysis, pneumonia, pleura effusion, fever, rigors, lethargy and anorexia.

The disease may reactivate as post-primary TB, exhibiting the initial symptoms or spreading to other organs and producing relative symptoms.

If the patient shows no symptoms, treat with isoniazid and rifampicin for 3 months. Symptomatic patients require complex treatments involving multiple antibiotics over an extended period.

TB worldwide, is becoming more virulent and resistant to antibiotics, immunisation is a far better than treatment.

Typhoid

Typhoid is caused by a faeco-oral contamination usually through consumption of poorly cooked food or dairy products. Symptoms include headache, rigors, fever, decreased appetite, constipation or diarrhoea, backache, nosebleed, tender abdomen.

Incubation can be anything between 3 days and 3 weeks. Fever rises daily for 7-10 days, peaks for 7-20 days then drops over the next 7-10 days. In the first stage of the disease, rose-coloured marks appear on the patient.

Treat symptomatic diarrhoea, dehydration, pain, fever and give Ciprofloxacin 250mg-500mg BD for 7-14 days.

Chapter 15 Poisoning

Poisoning after ingestion of Mushrooms Plants or Contaminated Water

If in a situation where food is scare you may be tempted to try and forage for wild plants and Mushrooms. In the case of mushrooms I would strongly advise not to try any you cannot positively identify as being safe. Mushrooms have very little food value and are not worth the risk.

In the UK Fatal Mushroom poisoning is rare but some species can cause unpleasant diarrhoea and vomiting (D&V), which in itself can lead to dehydration and hypovolemic shock through loss of fluids.

As a general rule the time between ingestion and onset of symptoms gives a clue to the seriousness. Interestingly those that cause symptoms early (within 6 hours) are less likely to be serious than those with a delayed onset. Although anyone who is seriously ill with lasting D&V should be considered a serious case.

Two particular poisonous mushrooms, *Amanita phalloides* (Death Cap) *and Amanita virosa* (destroying Angel) both contain a poison called amatoxin. These are pictured below;

Other common species can cause hallucinations, D&V and abdominal pain.

Poisonous plants such as belladonna, yew and elder all have different effects in addition to possibly causing D&V. In all cases the patient should be treated for the symptoms they exhibit as blood tests and other hospital investigations will not be available.

Always filter, boil and purify water from wild sources where possible.

Treatment for poisoning

Activated charcoal can be used within an hour of ingestion of most plant based poisons although its effectiveness is uncertain in most cases.

In any patients displaying symptoms monitor pulse, Blood pressure and GCS (see appendix 1). If the equipment is available and you understand it monitor cardiac rhythm and Perform a 12 lead ECG.

Try and keep anyone suffering hallucinations calm, a dark environment may help, constantly reassure. Sedation should be avoided if you can manage without it.

Rehydrate the patient using clean water or preferably rehydration fluids these can be pre-bought in sachets or mixed yourself by dissolving 6 level teaspoons of sugar and 1/2 level teaspoon of salt in 1 litre of clean water.

If dehydration looks likely establish IV access to replace fluids (see Clinical skills). Avoid using anti-diarrhoea and anti-sickness medication if possible as the body uses these methods to rid the body of toxins. Bouts of Diarrhoea usually stop on their own within 12 to 24 hours. If diarrhoea is severe and persists it could have been caused by bacteria ether from untreated water or surface contamination on food.

As different organisms can cause the infection it may not respond to a particular antibiotic. The recommended order to try them is;

Ciprofloxacin 500mg x2 first dose, then one tablet twice a day for three days.

Azithromycin 500mg for three days.

Metronidazole 600mg, three times a day for 8 days

If no improvement antibiotics can be changed after two days on each course.

Other rarer symptoms particularly from fungi can include;

Convulsions (See treatment for Epilepsy)

Decrease Respiratory Effort (See Resuscitation)

Chapter 16 Shock

Shock (Recognition, Prevention & Treatment

If you've ever injured yourself in addition to the pain you may have experienced being weak, dizzy, and nauseous, in which case you have experienced a mild form of shock. In this case, the symptoms appeared immediately after the injury, but they may not show up for several hours. Shock is a condition in which blood circulation is seriously disturbed. Crushed or fractured bones, burns, prolonged bleeding, asphyxia and dehydration can all cause Shock. Shock may be slight or it may be severe enough to be fatal. Because all traumatic injuries result in some form of shock, you should learn its symptoms and know how to treat the casualty. The best approach to **Shock Prevention** is to treat all casualties suffering from moderate and severe injuries for shock even if they are no showing immediate signs or symptoms.

Shock is frequently the most serious consequence of an injury

Types of Shock

Shock can be caused by several different causes:

• Hemorrhagic shock - reduced blood volume caused by either external or internal bleeding.

• Intestinal obstruction - results in the movement of large amount of plasma from the blood into the intestine.

• Severe burns - loss of large amounts of plasma from the burned surface.

• Dehydration - results from severe and prolonged shortage of water intake.

• Severe diarrhoea or vomiting - loss of plasma through the intestinal wall.

• Neurogenic shock - rapid loss of vasomotor tone that leads to vasodilatation to the extent that a severe decrease in blood pressure results.

• Brain damage - leads to an ineffective control of the brain's ability to constrict and dilate blood vessels (*medullary vasomotor*).

• Anaesthesia - deep general anaesthesia or spinal anaesthesia that decreases the activity of the area of the brain that controls constriction and dilation of blood vessels (*medullary vasomotor*).

• Emotional shock (Faint) - results from emotions that cause strong parasympathetic stimulation of the heart and results in vasodilatation in skeletal muscles and in the viscera.

• Anaphylactic shock - results from an allergic response that causes the release of inflammatory substances that increase vasodilatation and capillary permeability.

• Septic shock or "blood poisoning" - results from peritoneal, systemic, and gangrenous infections that cause the release of toxic substances into the circulatory system, depressing the activity of the heart, leading to vasodilatation, and increasing capillary permeability.

• Cardiogenic shock - occurs when the heart stops pumping or performance is decreased in response to conditions such as heart attack or electrocution.

How to recognise Shock

A person who is going into shock may show quite a few different signs or symptoms, some of which are indicated the diagram below and are discussed in the following paragraphs. Remember that some signs of shock do not always appear at the time of the injury; and, in some very serious cases the symptoms may not appear until hours later. Shock is caused directly or indirectly, by the disturbance of the circulation of the blood. Symptoms of shock include the following:

- The pulse is weak and rapid.

- Breathing is likely to be shallow, rapid, and irregular, because the poor circulation of the blood affects the breathing centre in the brain.

- The temperature near the surface of the body is lowered because of the poor blood flow; so the face, arms, and legs feel cold to the touch.

- Sweating is likely to be very noticeable.

- A person in shock is usually very pale, but, in some cases, the skin may have a bluish or reddish colour. In the case of victims with dark skin, you may have to rely primarily on the colour of the mucous membranes on the inside of the mouth or under the eyelid or under the nail bed. A person in or going into shock has a bluish colour to these membranes instead of a healthy pink.

- The pupils of the eyes are usually dilated (enlarged).

- If the casualty is conscious, they may complain of thirst. They may have a feeling of weakness, faintness, or dizziness, or they may feel nauseous.

- The casualty may be very restless and feel frightened and anxious. As shock deepens, these signs gradually disappear and the victim becomes increasingly unresponsive to what is going on around them. Even pain may not arouse them. Finally, the victim may become deeply unconscious.

© Lev Olkha | Dreamstime.com

You are unlikely to see all the symptoms of shock in any one case. Some symptoms may appear only in later stages of shock when the disturbance of the blood flow has become so great that the person's life is in serious danger. Sometimes other signs of the injury may disguise the signs of shock. You must recognise which symptoms indicate the presence of shock, but don't ever wait for symptoms to develop before beginning the treatment for shock.

Remember, every seriously injured person is likely to develop serious shock!

Prevention and Treatment of Shock

You should begin treatment for shock as soon as possible. Prompt treatment may prevent the onset of shock or, if it has already developed, prevent it reaching a critical point. Keep the victim lying down and warm. If conscious, the victim should be encouraged and assured that expert medical help will arrive soon.

Keep an injured person warm enough for comfort, but do not let the victim become overheated.

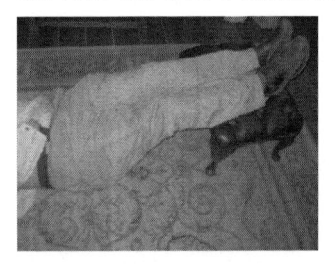

The best position to use to reduce the onset or to help in the treatment of shock is one that encourages the flow of blood to the brain. If possible, place the injured person on their back on either the floor, a bed, cot, or stretcher. Raise the lower end of the support about 12 inches so that the feet are higher than the head. If this is not possible, raise the feet and legs enough to help the blood flow to the brain. Sometimes it may be possible to take advantage of a natural slope and place the victim so that the head is lower than the feet.

Of course every case is different and you will have to consider the type of injury before you can decide on the best position for the patient. Here are some examples:

- If a person has a chest wound, they may have so much trouble breathing that you will have to raise the head slightly.

- If the face is flushed, rather than pale, or if you have any reason to suspect a head injury, don't raise the feet. Instead, you should keep the head level with or slightly higher than the feet.

- If the person has broken bones, you will have to judge what position would be best both for the fractures and for shock. A fractured spine must be immobilised before the victim is moved at all, if further injuries are to be avoided.

If you have any doubts about the correct position to use, have the victim lie flat on their back. The basic position for treating shock is one in which the head is lower than the feet. Do the best you can under the particular circumstances, to get the injured person into this position. NEVER let a seriously injured person sit, stand, or walk around.

No particular harm will be done if you allow the victim to moisten their mouth and lips with cool water, if it will make them more comfortable. Administer liquids sparingly and not at all if medical attention will be available within a short time. If necessary, small amounts of warm water, tea, or coffee may be given to a victim who is conscious. Persons having serious burns are an exception. Burn victims require large amounts of fluids. Water, tea, fruit juices, and sugar water may be given freely to a victim who is conscious, able to swallow, and has no internal injuries. Slightly salted water is also beneficial. This should be done if they are fully conscious, able to swallow, and you do not suspect that they have suffered internal injuries. Never give alcohol to a person in shock, as this will cause blood vessels to dilate and reduce blood pressure.

An injured person may or may not be in pain. The amount of pain felt depends in part on the person's physical condition and the type of injury. Extreme pain, if not relieved, can increase the degree of shock. Make the victim as comfortable as possible. Fractures should be immobilised and supported. Immobilisation greatly reduces, and sometimes eliminates, pain.

An injured person's body heat must be conserved. Therefore, warmth is important in the treatment of shock. Exposure to cold, with resulting loss of body heat, can cause shock to develop or to become worse. You will have to judge the amount of covering to use by considering the weather and the general circumstances of the accident.

Often a light covering will be enough to keep the casualty comfortable. Wet clothing should be removed and dry covering provided, even on a hot day. Use blankets or any dry material to conserve body heat. Artificial means of warming (hot water bottles, heated bricks, heated sand) should not ordinarily be used. Artificial heat may cause loss of body fluids (by sweating, and it brings the blood closer to the surface, defeating the body's own efforts to supply blood to the vital organs and to the brain. Also, the warming agent may burn the victim.

The treatment of Shock. In many emergency situations, is the most helpful thing you can do for an injured person. If shock has not yet developed, the treatment may actually prevent its occurrence; if it has developed, you may be able to keep it from reaching a critical point.

Emotional Shock

The impact of this type of shock will vary widely. Comfort and reassurance coupled with rest and relaxation after you are clear of immediate dangers is very effective in management of the casualty suffering from emotional shock.

It is important to keep the victim as calm as possible because excitement and fright will affect their condition and may even bring on shock. Try to prevent the victim from seeing their injuries, and reassure them that they will be properly cared for. Keep all unnecessary persons away, as their conversation regarding the victim's injuries may increase their agitation.

Table 1, Clinical Signs of Shock

	Class 1	Class 2	Class 3	Class 4
Blood Loss Volume (mls) in adult	750mls	800 - 1500mls	1500 - 2000mls	>2000mls
Blood Loss % Circ. blood volume	<15%	15 - 30%	30 - 40%	>40%
Systolic Blood Pressure	No change	Normal	Reduced	Very low
Diastolic Blood Pressure	No change	Raised	Reduced	Very low / Unrecordable
Pulse (beats /min)	Slight tachycardia	100 - 120	120 (thready)	>120 (very thready)
Capillary Refill	Normal	Slow (>2s)	Slow (>2s)	Undetectable
Respiratory Rate	Normal	Normal	Raised (>20/min)	Raised (>20/min)
Urine Flow (mls/hr)	>30	20 - 30	10 - 20	0 - 10
Extremities	Normal	Pale	Pale	Pale & cold
Complexion	Normal	Pale	Pale	Ashen
Mental state	Alert, thirsty	Anxious or aggressive, thirsty	Anxious or aggressive or drowsy	Drowsy, confused or unconscious

Fluid replacement should only be given in sufficient volume to maintain a radial pulse (at the wrist) this indicates sufficient circulation to the extremities. Giving large volumes of fluid is detrimental as it interferes with the clotting process and dilutes the blood decreasing its ability to transport oxygen to vital organs.

Below are some fluid regimes that indicates the maximum amount of fluid that should be given, bearing in mind the points above.

Table 2. Suitable Blood Replacement Regimes for previously healthy adults

Estimated Blood loss	Suitable fluid regimes
1000 mls	3 L Saline or 1 L Haemaccel
1500 mls	(1.5 L Saline & 1L Haemaccel) or 4.5 L Saline
2000 mls	(1 L Saline, 1 L Haemaccel & 2 units blood or plasma) or (3 L Saline & 2 units blood or Plasma)

Chapter 17 Ear, Nose and Throat (ENT)

Ear Problems

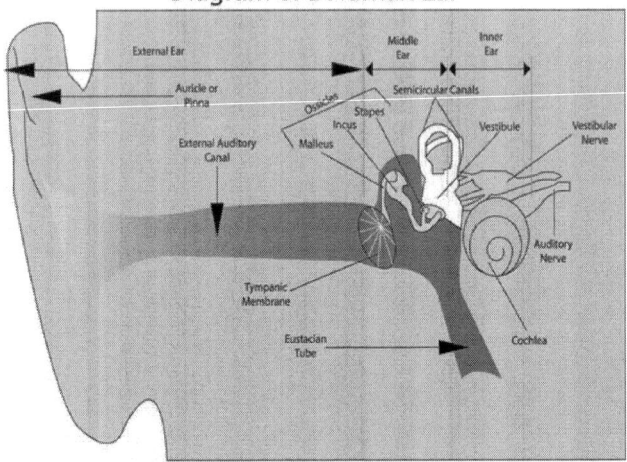

Diagram of a Human Ear

© Anita Potter – Dreamtime.com

Examination

Check hearing by whispering in each ear. Feel around ear for tenderness, wounds, swelling and ask about any pain.

I was once told the smallest thing that should be put into a ear is an elbow. So in other words don't stick anything blindly in as it can cause damage.

An instrument called an otoscope is used to examine the ears and sometimes the nose. It consists of a cone with a magnifying lens and a light source. With it you can look for infection, reddening, traumatic damage, pus and wax build up in the canal and to examine the ear drum for reddening or fluid levels.

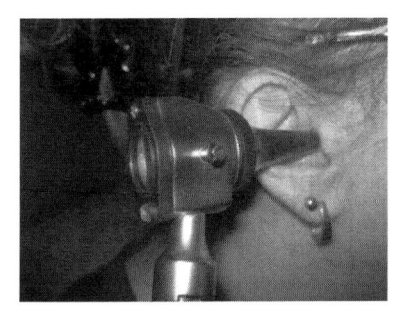

Foreign objects

Insects are the most common foreign objects found in the ear, if they crawl in, then bite, sting or die in the ear they can trigger an infection. The safest way of removal is to irrigate with oil or water and allow the object to float out. If this is not possible use grasping or magil`s forceps to get it out, care must be taken not to rupture the tympanic membrane 'ear drum'.

Rupture of the tympanic membrane

If this happens, treat the pain with a simple analgesia and cover the affected ear and avoid swimming. The ear will heal itself.

Infection of outer ear 'ottis externa'

Outer ear infections often occur in moist environments, such as with diving or expeditions in tropical regions. The auditory canal becomes red, itchy and there may be a discharge as well as hearing loss. Pulling on the ear lobe will be painful.

To treat, irrigate ear to soothe and remove debris and treat with a combination steroid and antibiotic cream such as Gentisone.

Infection of middle ear 'ottis media'

Often associated with an upper respiratory infection with the inside of the ear being painful and usually decreased hearing. If you examine the ear with an otoscope the inside will be reddened and pus may be seen behind the tympanic membrane.

Treatment is with analgesia and antibiotics such as amoxicillin or erythromycin.

Ear Wax

The presence of Ear wax can cause pain, but more so loss of hearing and a feeling of being underwater. Use drops to soften the wax this may allow it to run out on its own or ease the process of syringing them.

Nasal Problem

Congestion

Congestion is usually caused by allergies, if the cause is likely to be bacterial and the patient has a fever of >38 degrees give antibiotics, such as Doxycycline 100mg bd. Otherwise treat symptomatically and give decongestants, encourage the taking of oral fluids which will loosen the mucus build up.

Nose bleed

Treatment for a nose bleed is to get the patient to sit up and pinch the nose just below the bridge and hold for 10 minutes if it is still bleeding repeat and repeat again. If after 30mins it is still bleeding clear any clots, soak some ribbon gauze in adrenalin and pack into nose. This will apply direct pressure and the adrenalin will encourage the blood vessels to shrink which will help with clotting.

Other methods include simple packing with tampons or gauze or the use of a silver nitrate stick to cauterise the wound (apply for max of 5 seconds). For bleeding in the back of the nasal cavity apply direct pressure to the bleeding point using a balloon catheter which is inflated after pushing it into the nose.

Nasal Foreign Object

If a foreign object is known to be lodged in the nose, it needs to be removed as it may be inhaled causing breathing difficulties, swallowed or cause an infection. If the patient does not know there is something there, such as with an insect or a child who may have pushed an object up their nostril then it may already be infected and there will be a foul smelling discharge from the nose which will be accompanied by pain. Use a speculum or a thin nosed forceps or pliers to open up the nasal opening. It can be opened up fairly wide without causing any pain. Use a head torch or pen torch to look inside.

Use grasping or magil forceps to grasp any object to remove it. If the object has already broken up then it is better to wash the debris out. Use a syringe to irrigate through the other nostril and wash the piece out. A cotton bud can be used to assist with this but care must be taken that the object is not pushed further back. If the object does move into the back of the nasal cavity the patient may be able to cough it out.

Throat problems

The common cold doesn't require any treatment apart from symptomatic relief for congestion. Look for swelling off the tonsils and uvula. A lighted probe can be used with a mirror to examine the inside of mouth and back of throat. Whilst examining the throat look at the tongue it could be pink and moist.

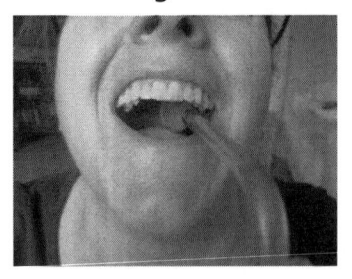

Tonsillitis

The tonsils are two small glands found at the back of the throat on each side. Tonsillitis is inflammation of the tonsils caused mostly by a viral or sometimes a bacterial infection.

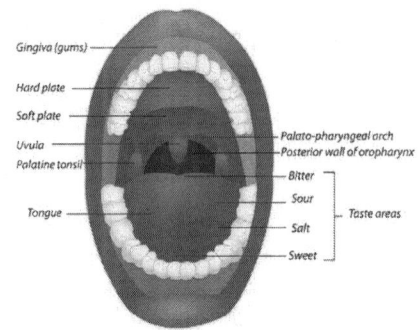

Guniita | Dreamstime.com

The main symptom of tonsillitis is a sore throat and pain when swallowing, there may also be coughing, headaches and a fever. White pus-filled spots will be visible on the tonsils and swollen glands (lymph nodes) in the neck.

The incubation is between 2 and 4 days. It is spread through droplets in the air, hands or on surfaces etc, anybody can get tonsillitis but it is most prevalent in 5-15 year olds.

Treatment

Normally tonsillitis clears up on its own in a few days, the patient needs to make sure they carry on eating and especially drinking although it may be painful to swallow.

Take paracetamol and/or ibuprofen to ease pain, headaches and fever. Lozenges can help with a sore throat

If the infection is bacterial it will respond to antibiotics but most are viral. Antibiotics only reduce infection by around a day and decrease risk of already rare complications such as rheumatic fever.

If antibiotic are deemed necessary, a 10 day course of Penicillin V is appropriate. If allergic to Penicillin use a 5 day course of erythromycin or azithromycin. If swelling is severe and infections continue to reappear, surgery is generally the answer, however this is not applicable to wilderness medicine.

Surgery

Tonsillectomy is performed using a general anaesthetic as the patient needs to be intubated. They are either removed with a tonsil scissors or tonsil knife then dissolvable sutures are inserted. They can also be removed by using Diathermy (Electric Heat) or Coblation (cold ablation). Bleeding is stopped either by direct pressure or sealing of the wound by the Diathermy or Coblation. Complete recovery takes about two weeks and person should be isolated to reduce risk of picking up infection in that time.

Chapter 18 Eye Problems

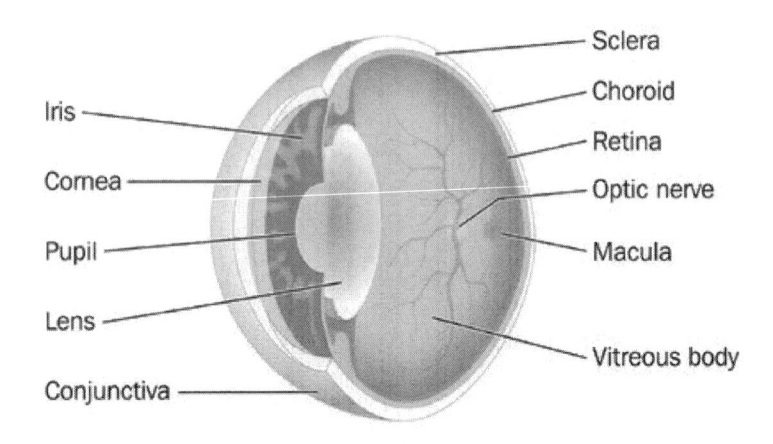

© Legger | Dreamstime.com

Examination

Most eye problems are either visible or give clear symptoms for a closer and more detailed eye exam an ophthalmoscope can be used.

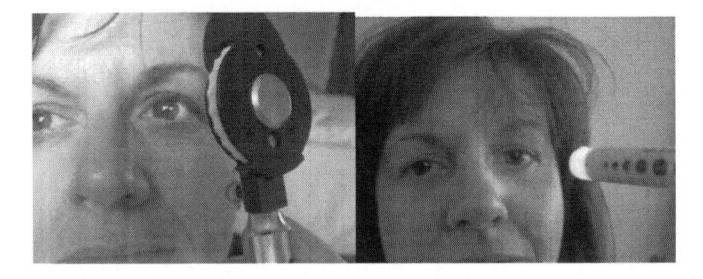

A pen torch with or without a magnifying glass could also be used.

Do the pupils react to light?

Check Vision is it normal for patient, compare both sides?

What do the pupils look like, size and shape?

What does the conjunctiva look like?

What does the front part of the eye look like?

Eye Infections

Patients with Eye Infections don't necessarily need them covered unless blinking irritates the eye. If secretions or pus is present these need to be wiped away. If both eyes are affected use separate cloths for each eye. There are a number of infections that effect the eye and surrounding tissues.

Conjunctivitis

Conjunctivitis is an infection of the surface of the eye, it feels like having grit in your eye. See below for different presentation.

Signs of Bacterial Infection;

- Widespread Reddening of the eye
- Pus
- Fever, Enlarged Lymph Nodes in neck

Signs of Viral Infection;

- Blotchy Reddening of the eye
- Watery excretions
- Fever, Enlarged Lymph Nodes in neck

Signs of Allergic Conjunctivitis;

- Reddening of the eye - Light pink in Colour
- Clear discharge
- Other allergy signs

Treatment

Both infections clear by themselves and only require irrigation with clean water. If they don't clear and a bacterial cause is suspected use Chloramphenicol ointment or drops as directed, if not available try an oral antibiotic such as Doxycycline 100mg twice a day. If an infection fails to clear recheck for a foreign body in the eye.

Iritis

A more serious condition is Iritis, inflammation of the iris. A red circle is seen around the edge of the iris. If you shine a pen torch in to the eye it doesn't constrict and will be painful. If untreated the pupil will become misshaped and the eye becomes cloudy. Treat with analgesia and corticosteroid drops or creams i.e those based on Dexamethasone or Bethamethasone and irrigate eye. Caution is advised as steroid eye preparations can cause glaucoma or cataracts.

Always examine to eliminate a foreign object in eye.

'Snow' Blindness

Snow blindness is caused by reflection of ultra-violet light off the ground, most often snow, but also water and desert conditions. Excessive ultra-violet exposure can also cause headache and cold sores. Any vision loss will heal itself usually within 72hrs but is very painful. Always examine to eliminate a foreign object in eye. Prevent Snow Blindness by wearing proper sunglasses which block ultra-violet light.

Treatment

- Give analgesia
- Rest in dark area
- Cover effected eyes
- Tropicamide 0.5% Drops will dilate the pupil.
- Amethocaine 0.5% will ease pain but delay healing
- Bathe frequently with cold water.
- Chloramphenicol 0.5% QDS will sooth and prevent infection

Eye Examination for foreign objects

If the eye is painful or feels gritty it should be checked for foreign objects. Perform a visual inspection of the visible surface, a magnifying glass may help. If any objects are visible they can usually be removed by using a damp cotton bud. If nothing is immediately obvious use a clean bud to roll up the top eyelid, this allows examination of both the top of the eye and inside of lid.

If an object is not removed it will form an ulcer on the surface and affect the patient's vision.

An opthalmoscope can be used to examine the eye which allows for a greater magnified view of the surface, but is often not required.

If the eye is very painful Amethocaine 0.5% drops can be used to dull sensation. Either pull down lower lid and administer one drop, if the eye is too painful too open, one or two drops can be placed in the corner of the eye when the patient is lying on their back. Then have the patient blink this distributes the drug over the surface of the eye. The drops initially hurt but the pain quickly subsides. Shine a pen torch in to the eye and often the foreign object will glint or cast a shadow, some may be very small. If the object is imbedded prod at side with cotton bud until it loosens then wipe away. Irrigation or blinking under water may also lift object.

It the object won't move and there is no access to professional health care then it is better to remove it than leave in place as it is likely to cause infection and permanent damage to sight.

In extreme circumstances if this fails the object can be removed by picking with a scalpel blade or needle. If removal is attempted, the eye will need to be treated beforehand with amethocaine 0.5% drops to dull sensation. If the object is so deeply imbedded leaving it for a few days will form an ulcer which may lift it from the eye surface, making it easier to remove. Although you run the risk of infection as stated above.

After any removal or foreign object treat with Chloramphenicoll 0.5% QDS to treat and prevent further infection and give oral analgesia.

Fluorescein 1% strips can be used after Amethocaine to dye the eye, which gives foreign objects a greater contrast and makes them easier to remove.

Corneal Abrasion

Corneal abrasion are small scratch on the surface of the eye, can be caused by grit, contact lens removal or other trauma. Always examine to eliminate foreign object in eye and treat with Chloramphenicol 0.5% QDS

Chapter 19 Skin Conditions

The skin is the largest organ of the body

- It Protects the body from infections
- Provides an interface by means of touch, sensation and temperature.
- Regulates your body temperature.

Cellulitis

This is an inflammation of the layers of the skin and is caused by a bacteria often introduced through open wounds, often affecting the face and lower legs. Cellulitis can be treated with antibiotics.

More superficial infections with defined boundaries can also occur and often coexist with Cellulitis.

Impetigo

Impetigo is a superficial skin condition causing blisters that when they erupt form yellow crusted lesions, that spread rapidly and are highly contagious. If the lesions are isolated treat with Fucidin cream, but if more widespread use oral antibiotics such as Flucloxacillin or Erythromycin.

Chapter 20 Minor Medical Problems

Constipation

Bowel habits vary enormously from person to person, and a change in habits and diet can provoke constipation. An adequate diet with fibre and sufficient fluids will often prevent or cure it. If pain and cramping becomes a problem increase fluids and give mild laxative. Do not over medicate as it may promote loose stools which is often unpleasant in an outdoors situation.

Diarrhoea

Diarrhoea is often a problem in wilderness settings and is usually brought on by a change of habits, food or drinking contaminated water. If no blood is present it is best to let it run its course and by giving the patient bland food and plenty of fluids, as this is the bodies reaction to try rid itself of a bacterial infection. If you need to travel take loperamide or similar medication to prevent the gut from going into spasm but avoid if possible. If in a relevant area and other symptoms are present consider the treatment for cholera or typhus.

Headache

Headache's have many causes; common ones include dehydration, heat exhaustion, tension, sinusitis, viral illnesses, migraine, lack of sleep and eye strain. Rarer causes include very high blood pressure bleeding within the skull, strokes, TIAs, concussion, hypoglycaemia, meningitis, pre-eclampsia and menstruation. If accompanied by a relevant history of trauma, pregnancy or with other symptoms treat accordingly. Most headaches though have an innocent cause and can be treated with rest in a cool, shaded location, simple analgesia and encouraging the taking of oral fluids.

Indigestion

Indigestion is often characterised by pain, nausea, belching, bringing up food or acid and a full feeling. Often self limiting goes away within the hour. If uncomfortable take an antacid for relief. In a number of cases people have thought they had indigestion but in fact were having a heart attack or vice versa.

Nausea and/or Vomiting

Should be treated with anti-emetic drugs mild attacks can be treated with oral swallowed medication where more severe cases can be treated tablets that dissolve in the mouth, injections or suppositories.

Chapter 21 Medication

Introduction

Within this section is information on common useful medication. Most are available only with a prescription, those freely available over the counter are marked (OTC), Those that can be dispensed by a Pharmacist are marked (P) and any Controlled Drugs are marked (CD) .

Pharmaceutical science is a huge subject, before any drug is taken or administered both parties need to have knowledge of its use, side effects, interactions and contra-indications. Patients with other existing conditions or who are taking other medication must be particularly careful when starting new ones. Both parties need also be vigilant in case serious side effects occur and know how to counter them.

It is beyond the scope of this book to go into detail on each drug, so information leaflets must be studied carefully and it is recommended you also purchase a reliable up to date Pharmacy Book, In the UK the British National Formulary (BNF) is published twice a year and is an excellent resource. If you ask you local chemists they may have some old ones they will give away.

Antibiotics

Example Dosages and Courses for using Antibiotics

If you get a prescription you get the correct dose put into a container if however you are stocking up you get tablets in Boxes or jars. The question is how many tablets will you need to provide a course against a particular infection. You can take multiple smaller tablets to make up a higher dose.

BD (Twice a day)

TDS (Three Times per Day)

QDS (Four Times per Day)

All doses are for adults see medication information leaflets for child doses.

Amoxicillin

Don`t administer if patient sensitive to Penicillin.

Bronchitis	250-500mg ever 8 hours
Pneumonia	500mg – 1g ever 8 hours
Dental abscess	3g repeat after 8 hours
Ottis Media	500mg – 1g ever 8 hours
Sinusitis	250-500mg ever 8 hours
Anthrax	500mg – 1g ever 8 hours

Azithromycin

Respiratory tract Infection	500mg OD for 3 days
Ottis Media	500mg OD for 3 days
Chlamydia	1g Single Dose
Urethritis	1g Single Dose
Typhoid	500mg OD for 7 days
Lyme Disease	500mg OD for 7-10 days

Benzylpenicillin

Don`t administer if patient sensitive to Penicillin

Pneumonia	IM /Slow IV 0.6g–1.2g QDS
Ottis Media	IM /Slow IV 0.6g–1.2g QDS
Endocarditis	1.2-2.4g ever 4 hours
Meningococcal Septicaemia	2.4g ever 4 hours
Throat Infections	IM /Slow IV 0.6g–1.2g QDS
Cellulitis	IM /Slow IV 0.6g–1.2g QDS
Limb Amputation	IM /Slow IV 0.6g–1.2g QDS
Anthrax	2.4g ever 4 hours

Cefotaxime

Gonorrhoea	1g BD
Surgical Prophylaxis	1g BD
Epiglottises	1g BD
Meningitis	2g ever 6 hours

Chloramphenicol

Meningitis	25mg/Kg QDS
Septicaemia	25mg/Kg QDS
Epiglottises	25mg/Kg QDS
Typhoid	12mg/Kg QDS

Eye Drops 0.5% 10ml (OTC)
Eye Ointment 1% 4g (OTC)
Ear Drops 5% & 10 % 2-3 drops TDS

Ciprofloxacin

Salmonella	500-750mg BD
Typhoid	500-750mg BD
Campylobacter	500-750mg BD
Shigellosis	500-750mg BD
Peritonitis	500-750mg BD
Pyelonephritis	500-750mg BD
Gonorrhoea	500mg Single Dose
Chlamydia	500-750mg BD
Gastro-intestinal Infection	500-750mg BD
Bone/joint Infection	500-750mg BD
Septicaemia	500-750mg BD
Anthrax	500mg BD
Surgical Prophylaxis	750mg 1hr before
Prostatitis	500mg BD for 28 days
Respiratory Tract	500mg BD
UTI	250mg BD

Clarithromycin

Skin / Soft Tissue	250mg BD 7days
Impetigo	250mg BD 7days
Cellulitis	250mg BD 7days
Animal Bites	250mg BD 7days
Ottis Media	250mg BD 7days
Lyme Disease	500mg BD 14-21 days
Respiratory Infection	250mg BD 7days

Pneumonia 500mg BD 14 days
Bronchitis 250mg BD 7days
Increase to 500mg BD 14days for Severe Infections

Clindamycin
Associated with antibiotic colitis, change if patient get diarrhoea

Osteomyelitis	150-300mg QDS
Abdominal Sepsis	150-300mg QDS
Erysipelas	150-300mg QDS
Cellulitis	150-300mg QDS
Septic Arthritis	150-300mg QDS
Peritonitis	150-300mg QDS
Falcipartum Malaria	150-300mg QDS
MRSA Associated	150-300mg QDS

450mg QDS for severe infections

Co-Amoxiclav

Animal Bites	375mg TDS 7days
Chest Infections	375mg TDS 7days
GI Infections	375mg TDS 7days
Abdominal Infections	375mg TDS 7days
Pneumonia	375mg TDS 7days
Dental Infections	375mg TDS 7days
Sinusitis	375mg TDS 7days
Ear Infection	375mg TDS 7days
Impetigo	375mg TDS 7days

Cellulitis 375mg TDS 7days
In Severe Infections 625mg TDS 7days

Doxycycline

Prostatitis	100-200mg OD
Sinusitis	100-200mg OD
Syphilis - Early	100mg BD 14 days
Syphilis - Late	100mg BD 28 days
Pelvic Inflammatory Disease	100mg BD 14 days
Chlamydia	100mg BD 7 days
Urethritis	100-200mg OD
Lyme Disease	100mg BD 10-14 days
Anthrax	100mg BD
Malaria	100mg OD
Peridontitis	100-200mg OD
Oral Herpes Simplex	100-200mg OD
Rosacea	100-200mg OD
Acne	100-200mg OD
Urinary tract infection	100-200mg OD

Erythromycin

Oral Infections	250-500mg QDS
Campylobacter	250-500mg QDS
Syphilis - Early	500mg QDS 14 days
Chlamydia	500mg BD 14 days
Lyme disease	500mg QDS 14-21 days

Urethritis	250-500mg QDS
Respiratory Tract	250-500mg QDS
Legionnaires Disease	250-500mg QDS
Skin Infections	250-500mg QDS
Prostatitis	250-500mg QDS
Prophylaxis Diphtheria	250-500mg QDS
Streptococcal	250-500mg QDS
Whooping Cough	250-500mg QDS
Acne	500mg BD
Rosacea	500mg BD 6-12 weeks

Flucloxacillin

Ottis Externa	250-500mg QDS
Pneumonia	250-500mg QDS
Impetigo	250-500mg QDS
Cellulitis	250-500mg QDS
Osteomyelitis	IV 2g QDS
Endocarditis	IV 2g QDS

Metronidazole

Anaerobic/Dental	500mg TDS 7 days
Clostridium difficile	500mg TDS 10-14 days
Leg Ulcers	400mg TDS 7days
Pressure Sores	400mg TDS 7days
Bacterial Vaginosis	400-500mg BD 5-7 days
Pelvic Inflammatory	400mg BD 14days

Disease	
Gingivitis	200-250mg TDS 3 days
Oral Infection	200mg TDS 3-7 days

Trimethoprim

Urinary Tract Infection	200mg BD
Bronchitis	200mg BD
Pneumonia	200mg BD

Administration of prescription-only medicines by lay people

According to section 1.2.3 of the Medicines for Human Use Act

Administration of prescription-only medicines
The legislation provides that no one may administer a parental prescription-only medicine otherwise than to himself, unless he is a practitioner or is acting in accordance with the directions of a practitioner.

The following list of medicines for use by parental administration, are exempt from this restriction when administered for the purpose of saving life in an emergency.

Adrenaline injection (1 in 1000) (Allergic emergencies)

Atropine sulphate injection (antidote to organophosphorous poisoning / Increases Heart Beat)

Atropine sulphate and obidoxime chloride injection (antidote to Nerve Agents)

Atropine sulphate and pralidoxime chloride injection (antidote to Nerve Agents-tabun)

Atropine sulphate, pralidoxime mesilate and avizafone injection (Antidote organophosphorus or nerve agent)

Chlorphenamine injection (Antihistamines)

Dicobalt edetate injection* (Antidote to Cyanide Poisoning)

Glucagon injection (Treatment of hypoglycaemia)

Glucose injection 50%* (Treatment of hypoglycaemia)

Hydrocortisone injection (Treatment of Anaphylactic Shock)

Naloxone hydrochloride* (Antidote for Opiod Poisoning)

Pralidoxime chloride injection (Antidote organophosphorus or nerve agent)

Pralidoxime mesilate injection (Antidote organophosphorus or nerve agent)

Promethazine hydrochloride injection (Antihistamines)

Snake venom antiserum

Sodium nitrite injection* (Antidote to Cyanide Poisoning)

Sodium thiosulphate injection (Antidote to Cyanide Poisoning)

Sterile pralidoxime (Antidote organophosphorus or nerve agent)

Pain Management

Pain is a primitive and survival symptom, it's the way our bodies tell us there is something wrong. Webster dictionary defines pain "as a basic bodily sensation induced by a noxious stimulus, received by naked nerve endings, characterized by physical discomfort"

Treatment approaches to pain include pharmacologic measures, detailed below other interventional procedures, physical therapy, physical exercise, application of ice and/or heat, and psychological measures, such as biofeedback and cognitive behavioural therapy.

Pain is either acute or chronic, acute pain has a sudden onset for a limited time and disappears when the cause is treated. Chronic pain starts as acute pain but lasts longer that is expected.

In acute pain, attention is focused on treating the cause of pain whereas in chronic pain, its concerned with reducing pain to give relief, limit disability and improve functionality.

Chronic pain comes in many forms the main ones being Nociceptive pain arising from damage to tissues and Neuropathic pain which affects either the central or peripheral nervous system. Sometime conditions such as Sciatica have both forms.

In chronic pain management, pain signals continue to be generated in the damaged area after it has healed which gives rise to chronic pain.

Analgesic(Pain killing drugs)

Analgesics are divided into different types of drugs, The choice of drug is dependent on the type and severity of the pain, the World Health Organisation (WHO) suggests a three step pain ladder for Nociceptive pain.

Mild Pain; Paracetamol or Ibuprofen

Moderate Pain; Co-codamol, Dihydrocodeine, Paracetamol/Ibuprofen combination

Severe Pain; Tramadol, Morphine, Fentanyl, Methadone, Pethidine

Analgesic for Nociceptive pain;

Non steroidal anti-inflammatory drugs (NSAIDS)

Mainly effective for nociceptive pain

Paracetamol, ibuprofen, naproxen, diclofenac and salicylates

Side effects of these drugs are, gastritis, worsening of asthma and kidney damage.

Opioids

Opioids are effective against all types of Pain

Morphine, codeine, fentanyl, meperidine, tramadol, pentazocine and propoxyphene

Side effects include addiction, respiratory depression and constipation.

Alpha 2 adrenergic drugs

Used to treat hypertension have sedative properties used to treat pain and anxiety. clonidine and tizanidine. Side effects include fatigue and dry mouth.

Steroids

To relieve the pain of arthritis and treat breathing problems. Side effects can include diabetes, osteoporosis

Antidepressants,

Elevates mood but also improves the physical functioning.

Tricyclic antidepressants (amitryptyline),

Selective serotonin reuptake inhibitors (fluoxetine).

Sleeping Pills and Muscle Relaxants

For pain stiffness and muscle spasm, mild sedative properties.

Carisoprodol, Methocarbomol and Diazepam.

Analgesic for neuropathic pain;

Opioids

Opioids are effective against all types of Pain

e.g. Morphine, codeine, fentanyl, meperidine, tramadol, pentazocine and propoxyphene

Side effects include addiction, respiratory depression and constipation.

Anticonvulsants

Designed for treatment of seizures but now also used for neuropathic pain.

carbamazepine, clonazepam, valproate, phenytoin, gabapentin, topiramate and lamotrigine

Local anaesthetics

Local anaesthetics given by mouth are useful for neuropathic pain. The most commonly used one is mexiletine which was originally used for heart rhythm abnormalities. Other drugs used are tocainide and flecanide.

Neuroleptics are drugs traditionally used for psychotic illnesses. Two drugs of this class olanzapine and resperidone are found useful to treat chronic pain.

Topical analgesics

These includes drugs like menthol, lignocaine, EMLA cream and capsaicin.

NMDA antagonists

NMDA receptors worsen pain. NMDA receptor antagonists therefore can relieve neuropathic pain. Some of the drugs of this class are methadone, dextromethorphan and ketamine.

Other Pain Relief techniques;

Transcutaneous electrical nerve stimulation (TENS)

TENS is a non-invasive procedure, intended to reduce pain by nerve stimulation, it works on both acute and chronic pain, its effective for postoperative pain, osteoarthritis, and chronic musculoskeletal pain particular back pain.

Ice or cooling

Ice works muscle strains and sprains, particularly in the first 48-72 hours following the injury. Ice reduces swelling and provide some pain relief by numbing surrounding tissue. Cooling encourages warm blood to the injury site, brining oxygen and nutrients to aid in the healing process. Never apply ice directly to the skin and no more than 20 minutes for a time. Cooling can be achieved with a chemical ice pack or a bag of ice wrapped in a towel

Heat or Warming

The application of Heat increases circulation to the injury site. It should be applied after 48-72 hours following an injury. Heat can be applied via a chemical pad or a moist hot towel. Be careful not to burn or scald the skin.

Individual Analgesics

Aspirin (OTC)

Suitable for mild pain, also used as an anti-inflammatory and to reduce temperature in fever. Sold in UK in 300mg and 75mg doses, Higher for general acute use lower given as a daily anti-platelet drug to prevent stroke and heart attacks. Recent research indicates also reduces chance of getting some cancers.

Co-Codamol (OTC & POM)

A combination of codeine phosphate and paracetamol. Co-codamol is used for the mild to severe pain. In the UK two strength 15mg Codeine/500mg Paracetamol and 30mg Codeine/500mg Paracetamol tablets are available only with a prescription, and 8mg Codeine/500mg Paracetamol and 12.8mg Codeine/500mg Paracetamol strength is available without a prescription.

Codeine (POM)

Is a mild opiate used for Mild to moderate pain, as a cough medicine and in treating diarrhoea as one of the side effects is constipation

Diclofenac (POM)

Is a non-steroidal anti-inflammatory drug (NSAID) taken to reduce, fever, inflammation and reducing pain in conditions such as arthritis or acute injury.

Dihydrocodeine (POM)

Is an opiate used moderately severe pain as well as coughing and shortness of breath, either alone or mixed with aspirin or paracetamol. In the UK, 30 mg tablets containing only dihydrocodeine is available, also a 40mg Dihydrocodeine tablet is available called DF-118.

Entonox (P)

A mix of 50% Oxygen and 50% nitrous oxide is both an analgesic and anaesthetic gas. Used in dentistry, child birth and emergency medicine. It is fast acting taking around 30 seconds to take effect but effect also wears off in about 60 seconds so constant use is recommended. It's self administered and the patient must inhale deeply to operate the demand valve. Shouldn't be used in patients with suspected punctured lung and used with caution on patients who are intoxicated or with head injuries as it will mask any deterioration in level of consciousness.

Ibuprofen (OTC & POM)

Is a non-steroidal anti-inflammatory drug (NSAID) used for mild pain, available as 200mg or 400mg tablets. Originally used for the treatment of arthritis.

Methadone (POM)

Is a opiate used mainly for treating opiate dependency, but is an analgesic and cough suppressant also used in the long term treatment of chronic Pain.

Morphine (POM)

Is the best medication to treat severe pain, available as tablet, suspension and injection. Should be used with caution as it is a respiratory and cardiac depressant. These effects can be counteracted with Narcan

Tramadol (POM)

Is used in treating moderate to severe pain. Available as tablet or injection.

Paracetamol (OTC)

Used to treat mild pain and reduce fever, very useful for headaches, colds and flu, Available as tablet, suspension, injection and suppository

Pethidine (POM)

Is for moderate to severe pain, can be given as a tablet, suspension or by injection. It used to be very popular particularly for surgical problems. But less so in recent years as Morphine has been found to be as effective with fewer side effects.

Naloxone Hydrochloride (Narcan) (POM)

Narcan is used to counter the effects of heroin or morphine overdose. Can be given IV, IM or nasally in 400mcg doses. Maximum dose is 10mg

Heart (Cardiac) Drug Management

Different types of Drugs are available to reduce the risk of new or recurring Cardiovascular disease. Always seek professional advice before commencing any drug treatment.

Anti-platelet drugs

The most common Anti-platelet drug is Aspirin 75mg once per day, Aspirin should be used cautiously with asthma, ulcers and High Blood Pressure. A second anti-platelet drug is clopidogrel 75mg.

Lipid-Lowering Drugs

If Cholesterol is a problem Simvastatin 10 mg can be bought without a prescription. Branded as Zocor Heart plus. See below for £6.25 or Boots Chemist for £7.99 a pack.

If you have a moderate risk factor the usual starting dose is 20mg which is initially reviewed after 3 months then regularly at 6 or 12 month intervals.

Anti-Coagulant Drugs

Anti-coagulant Drugs most commonly Warfarin as are prescribed to patients with Chronic Heart Conditions or Atrial fibrillation (a conductive problem with the heart) with a history of Strokes, TIAs or over 65 years. Dosages of Warfarin are given depending on a measurement of the clotting time called INR, this requires a blood sample and laboratory test to manage accurately.

Anti-Hypertensive drugs

Treatment should be started in all people if BP is consistently >160/100.

If at high risk of Cardiovascular Disease (CVD) or History of CVD, Stroke or Diabetes. Treatment threshold is consistently >140/90

There are many drugs that are used in the treatment of Hypertension by different actions all reduce blood pressure and many have other beneficial cardiac affects. Below are some examples of commonly prescribed ones.

Current guidelines suggest that for patients over 55 the best drug treatment is using either a calcium-channel blocker such as (Verapamil 80mg TDS or Nifedipine/Adalat 30mg OD) or a thiazide-type diuretic such as Bendroflumethiazide 2.5mg OD.

For patients under 55 use an ACE inhibitor such as Captopril 12.5mg BD or an angiotensin-II receptor antagonist sich as Eprosartan 600mg OD

Anti-Anginal Drugs

Glyceryl Trinitrate

Used to treat angina and for symptomatic relief in Left ventricular Failure(LVF) and Heart Attacks (Myocardial Infarction(MI)). it comes in several forms as spray, patch, ointment or tablet which dissolves in the mouth. The main action of the drug is to make blood vessels bigger (a vasodilator) this results in a decrease in blood pressure and less work for the heart. This is intended to reduce the patients pain.

Should br cautiously given to patients who already have low blood pressure < 90mmHg. Those in Hypovolaemic (low fluid) shock. Anyone who is suspected to have bleeding in their head or with a Head injury and anyone unconscious or who has taken Viagra.

The Side Effect is that it increases the size of vessels it causes a headache, dizziness and the patient may faint if sat up or stood suddenly (postural Hypotension).

Other Anti-Anginals include Isosorbide Dinitrate and Isosorbide Mononitrate.

Diuretic Drugs

Diuretics are used to remove excess fluid held in the lungs and other tissues caused by chronic Heart failure. Examples includes Frusemide and Bendroflumethiazide. The minimum dose required for effect should be given. One side effect is the loss of potassium from the body which will need to be replaced as a low potassium level can cause the heart rhythm to change. A dose of Frusemide 40mgs can be given intravenously in an emergency after GTN to relieve pulmonary oedema in Heart failure.

Thrombolytic Drugs

Thrombolytic drugs are given to dissolve clots within the body. Primarily used in prehospital care to dissolve clots in Heart Attacks the same drugs such as Reteplase or Tenecteplase are used in the treatment of Clots in the Lungs (Pulmonary Embolisms) or in the Brain (Cerebral Embolisms) as caused by thrombolytic Strokes. There is comprehensive checklist for each type of condition. The most important contra-indications are where the patient is taking Warfarin or other blood thinners or if they have had recent surgery as there is a high risk of bleeding.

Respiratory Drug Management

Salbutamol (Ventolin)

Relaxes smooth muscles in airways, which spasm in acute asthma, effect increased if nebulised with oxygen. Given in Asthma attacks. COPD and Anaphylaxis. Normally taken as Aerosol (100-200mcg 1-2 puffs), inhaled powder (200-400mcg) or nebulised solution(2.5mg-5mg), but also available as tablet and inject able solution.

Ipratropium Bromide (Atrovent)

Atrovent is given secondary to Nebulised salbutamol in Life threatening Asthma, Acute COPD. It acts as a long acting bronchodilator. It can be given along with the first dose of salbutamol or with subsequent doses if first dose of salbutamol is not successful. Dose is 500mcgs of nebulised Atrovent.

Adrenaline/Epinephrine

Adrenaline can be used in the emergency treatment of acute anaphylaxis and life threatening Asthma. Given by intramuscular injection of 0.5mg in outer thigh or in upper arm.

Hydrocortisone

Is used in the emergency treatment of acute anaphylaxis and life threatening Asthma. Given by intramuscular or slow intravenous injection of 200mg hydrocortisone over two minutes.

Drugs used in Cardiac Arrest

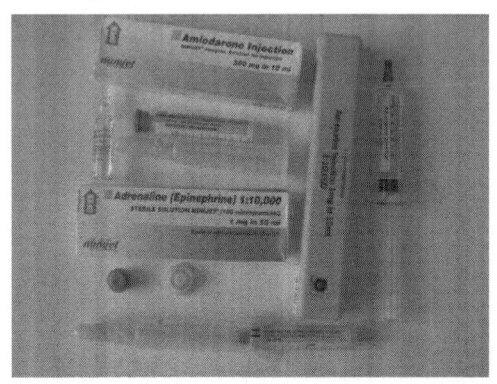

Drugs used in Cardiac arrest come in ampoules, prefilled syringes or minijets which consist of a vial of medication and a syringe applicator which is designed to attach to the giving port of a standard cannula for IV administration.

If the patient is in Ventricular Fibrillation (VF) or Ventricular Tachycardia (VT). Start 1mg adrenaline after delivering the third shock and then repeat ever 3-5 minutes. Also give amiodarone after the third shock. If the patient is in a non-shockable rhythm give 1mg adrenaline IV as soon as possible then repeat ever 3-5 minutes.

Diabetic Drug Management

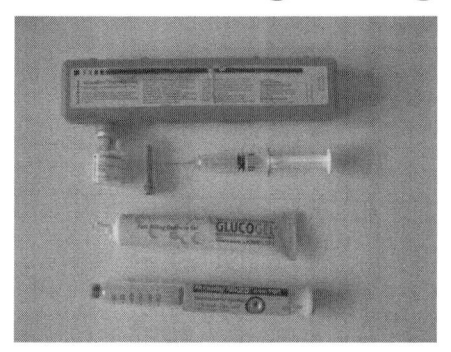

Insulin

Balancing Insulin regimes can be complicated, Insulin comes in short, intermediate and Long acting varieties. It must be taken to balance both the sugars taking into the body and the exercise done. It is usually given by injections under the skin. Some patients have a pump which gives a constant supply of short acting insulin.

Hypostop/Dextrose/Gluco Gel

A tube containing a high concentration 40% Glucose Gel, can be given to conscious casualties to rubbed into gums of unconscious casualties providing they are in the recovery position and wont choke on it.

Glucagon / Glucagen

This is an emergency drug provided as a powder and a syringe of sterile water to reconstitute it. This is given as an intramuscular (IM) injection. It converts glycogen to glucose in the liver and raises blood sugar levels

Metformin

This is often the first medicine that is advised for type 2 diabetes. It mainly works by reducing the amount of glucose that your liver releases into the bloodstream.

Gliclazide

Increase the amount of insulin produced by your pancreas. They also make your body's cells more sensitive to insulin so that more glucose is taken up from the blood.

Acarbose

This slows down the absorption of carbohydrate from the stomach and digestive tract, preventing a high peak in the blood glucose level after eating a meal.

Nateglinide and Repaglinide

These stimulate the release of insulin by the pancreas. They are not commonly used but are an option if other medicines do not control blood glucose levels.

Thiazolidinediones (glitazones)

(e.g. pioglitazone) These make the body's cells more sensitive to insulin so that more glucose is taken up from the blood. They are a third line treatment for people who do not respond to other treatments or in whom other treatments are not suitable.

Gastric Drug Management

Anti-Emetics

These are used to treat nausea to treat nausea and vomiting. They are available as tablets, suspensions, suppositories, Injections etc. A useful buccal preparation is Buccastem which contains the drug Prochlorperazine which dissolves in the mouth thus doesn't need to be swallowed where active vomiting is present. Other useful drugs in this group are Cyclizine, Metoclopramide and Domperidone.

Anti-Diarrhoeal

Bacterial Diarrhoea should usually be allowed to run its cause but if you have to travel or are low on replacement fluid the medication is advised. Loperamide is the most common drug to use. Some people with long term bowel problems also take codeine phosphate as it has similar effect, although long term use often causes painful chronic constipation.

Laxatives

Laxatives work in different ways either increasing the fibre content of the gut or making it work harder to push out stools. Drugs are also available to soften stools and clean out the bowel depending on the desired effect. Commonly used examples are Fybogel, Lactulose, Movicol, Magnesium salts and Picolax.

Other useful medications include;

Anusol for Haemorrhoids.

Ranitidine, Peptac and gaviscon for indigestion.

Omeprazole and Lansoprazole for reducing gastric acid where ulcers are present.

Altitude Drug Management

Three main drugs are used in the treatment of Acute Mountain Sickness, High Altitude Pulmonary Oedema and Cerebral Oedema. See High Altitude Conditions for further details. The drugs are Acetazolamide, Dexamethasone and Nifedipine.

Other Useful Medication

AntiHistamines

Antihistamines are a used to treat allergies, hayfever, skin conditions and conjunctivitis. They work by blocking the effects of the protein histamine. Two types exist sedating and non-sedating, examples include Chlorphenamine, loratadine and cetirizine.

Diazepam

Diazepam is a sedative you to treat depression. In emergency medicine it is used in the treatment of seizures. Normally administered IV or rectally

Synometrine

Synometrine is given after the birth of a baby, it helps the delivery of the placenta and reducing risks from vaginal bleeding by causing the blood vessels to narrow (vasoconstriction).

Chapter 22 Vaccines (Immunisation)

The following information is taken from the UK NHS website and shows recommended vaccines based on age and exposure risk. Most vaccines are given as intramuscular injections. If administering vaccines be prepared for anaphylactic reactions and read vaccine literature to make recipient aware of possible side effects.

Children (Routine)

DTaP/IPV/Hib or 5-in-1 vaccine
Protects against: diphtheria, tetanus, pertussis (whooping cough), polio and Hib (haemophilus influenza type B). Given at: 2, 3 and 4 months of age.

Pneumococcal (PCV)
Protects against some types of pneumococcal infection. Given at: 2, 4 and 12-13 months of age.

Meningitis C (MenC)
Protects against meningitis C (meningococcal type C). Given at: 3 and 4 months of age.

Hib/MenC (booster)
Protects against haemophilus influenza type b (Hib) and meningitis C. Given at: 12-13 mnths of age.

MMR
Protects against: measles, mumps and rubella. Given at: 12-13 months and at 3 years and 4 months of age, or sometime thereafter.

DTaP/IPV (or dTaP/IPV) 'pre-school' booster
Protects against: diphtheria, tetanus, whooping cough and polio. Given at: 3 years and 4 months of age or shortly thereafter.

Children (Optional)
Varicella
Protects against: chickenpox.
Given: from one year of age upwards (one dose for children from one year to 12 years. Two doses given 4-8 weeks apart for children aged 13 years or older).

BCG (Bacillus Calmette-Guerrin)
Protects against: tuberculosis (TB). Given: from birth to 12 months of age.

Flu
Protects against: seasonal flu. Given: from six months and over in a single jab every year in Oct / Nov.

Swine flu
Protects against: swine flu. Who needs it: children with long-term health conditions or weakened immune system. Given: as part of the swine flu programme in 2009/10.

Hepatitis B
Protects against: hepatitis B.
Who needs it: children at high risk of exposure to hepatitis B, and babies born to infected mothers. Given: at any age, as four doses given over 12 months. A baby born to a mother infected with hepatitis B will be offered a dose at birth, one month of age, 2 months of age and one year of age.

Teenage (Routine)
Teenagers are advised to get the following if complete courses were not received as a child.

Teenage booster (Td/IPV)
Protects against: tetanus, diphtheria and polio.
Given at: between ages 13 and 18.

Cervical cancer vaccination (HPV, or human papillomavirus vaccination)
Protects against: human papillomavirus, which has been shown to cause cervical cancer in women.
Given at: 12-13 (girls only) and also, for the time being, to girls aged between 13 and 18 as part of a catch-up programme.

Teenagers (At Risk or missed in Childhood)
Meningitis (MenC) vaccine
MMR vaccine

Flu/Swine Flu vaccine
Hepatitis B vaccine

Adult (For at Risk groups)
Seasonal / Swine flu vaccine
Pneumococcal vaccine (PPV)
Varicella (chickenpox) vaccine
Hepatitis B (hep B) vaccine
BCG - Tuberculosis (TB)

Other Vaccines available for foreign travel
Typhoid, Cholera, Hepatitis A, Yellow Fever, Rabies, Japanese / tick-borne encephalitis, Meningitis, Plague

Chapter 23 Routes of Drug Administration

If you ask most people how you can take medication most will think of tablets, liquids such as cough medicine or injections n your arm, these are only three methods there are actually many more.

Auricular
Pertaining to the ear, may be a local anaesthetic to remove an abscess or drops in the ear to loosen wax, treat infections or relive headache, cold and/or flu symptoms.

Buccal
Pertaining to the cheeks or the mouth cavity, a tablet is placed n the cheek and the drug is then absorbed through the Buccal mucosa membrane. A useful method when the patent is unable to either swallow or is actively vomiting. Drugs such as midazolam for fits or anti-sickness (Anti-Emetic) medication can be given by this method.

Cutaneous
Pertaining to the skin, creams are applied cutaneously, Applies to anti-histamine, Analgesia, specialist burns creams and treatment for skin disorder.

Endocervical

Pertaining to the cervix, treatment options are generally limited to conditions relating to fertility, STDs and Tumors

Endosinusal

Pertaining to the sinuses, most of the medication given this way is for treating sinusitis or congestion and includes, nasal sprays, inhalers, drops and saline irrigation.

Endotracheopulmonary

Pertaining to the trachea and lungs, medication is given this way to an intubated patient. In hospital this is used to treat respiratory distress syndrome and pre-hospital in cardiac arrest if you are unable to obtain either vascular or intra-osseous access.

Epidural

An injection into the epidural space which inside the spinal canal but outside the sheath that protects the spinal card. An Epidural injection can be used in labour and provides both partial anaesthetic and analgesic effect.

Extra-amniotic

Administration of a drug in the space between the foetal membranes and endometrial inside the uterus. Used to induce labour.

Gastro enteral

A tube is past into the stomach either through the nose or mouth, this can then be used for feeding or medication. (See Clinical Skills)

Gingival

Applied to the gums applies to toothpaste or topical analgesia, antibiotics etc

Haemodialysis

Not Strictly a method of drug administration, used for removing water, creatinine and urea from the blood in cases of renal failure,

Intra-amniotic

Administration of a drug directly into the amniotic sac to effect the foetus.

Intra-arterial

Injection into an artery not common but occasionally used.

Intra-articular

Injection into cavity of joint used for pain management.

Intrabursal

Intrabursal injections are given into small sacks of fluids (bursae) between the tendons and bones

Intracardiac
Injections into the heart muscles or ventricles, rarely used despite what TV portrays

Intracervical
Into the Cervix, used in insemination etc

Intracoronary
Given directly into the coronary vascular system used with clot busting drugs

intradermal
Injecting between layers of skin for allergy testing

Intradiscal
Injecting between spinal discs for pain relief

Intralesional
Injecting into lesion for direct treatment with steroids etc

Intralymphatic
Injecting into lymph nodes or lymphatic system

Intramuscular
Injecting into muscles used for the slow absorption of drugs. (See Clinical Skills)

Intraocular
Injecting into the glob of the eye, used in treatment of ocular problems

Intraperitoneal
Injecting into peritoneal cavity used in human for treatment of cancers

Intrapleural
Injection into the pleura the lining of the lungs

Intraasternal
Into the sternum, a favourite place for bone guns by US medics

Intrathecal
Injection into the subarachnoid space of the spinal cord

Intraosseous
Injecting directly into the bone, used in Cardiac Arrest and in unconscious casualties where IV access is unobtainable (See Clinical Skills).

Intrauterine
Injection into the uterus used n terminations

Intravenous
Injection into the vein the most popular invasive measure (See Clinical Skills).

Intraventricular
Injections into the cerebral ventricles or subarachnoid space inside the skull.

Intravesical
Injection into the bladder

Nasal
Spray, aerosol, powders via the nasal passage

Ocular
Drops, irrigation in the eyes.

Oral
Via the mouth in solid or liquid form

Periarticular
Applied to a joint

Perineural
Applied to a nerve or group of nerves

Rectal
Absorbed though Rectal mucosa (see Rectal Diazepam for Seizures)

Respiratory
Inhaled / Nebulised Medication (See Clinical Skills)

Sub-conjunctival
Between the conjunctiva and the sclera of the eye

Subcutaneous
injection into the skin used for insulin, some inoculations, fluids and pain killers. (See Clinical Skills)

Sub-lingual
Absorbed through mucosa membrane under tongue used for rapid absorption for GTN

Transdermal
Absorbed through the skin via a patch e.g. Nicotine, GTN, Pain Killer etc

Urethal
Given through the urethra

Vaginal
Absorbed through Vagina

Chapter 24 Clinical Skills

Cannulation How and Why

The skill of cannulation needs to be practiced before you have to use it for real. A cannula is a hollow tube of plastic with a cap at one end and wings, which can be used to, attach it to the skin. It is usually covered by a custom made dressing that stops infection entering the insertion site, holds the cannula in place and provides a clear cover from which the site of insertion can be inspected for signs of infection and swelling. The cannula has a needle running through the centre which is used to puncture the skin and enter the vein. The needle is withdrawn after insertion leaving the plastic tube in the vein.

Cannulas come in various sizes as details below, the type you choose will depend on the circumstances. The table below shows the different types. A small cannula such as Pink (20G) or Blue (22G) will be easier to insert, but as it is smaller will allow less fluid to pass through in a given period. A good middle choice is a Green (18G) cannula. But where a casualty has lost a large volume of blood you will need two Orange or Brown (14G) cannulas to replace the blood volume quickly if required.

Cannula Sizes

Uses	Colour	Size	Maximum Flow rate
Rapid Transfusions / Trauma / GI Bleed	Brown/ Orange	14G	275 mls/min
Rapid Transfusions / Obstetrics	Grey	16G	170 –180 mls/min
Transfusions / Cardiac / Neuro	Green	18G	80 – 100 mls/min
Infusions / Bolus Medication	Pink	20G	54 – 60 mls/min
Bolus Medication	Blue	22G	25 – 30 mls/min
Bolus Medication / Short Term Infusion	Yellow	24G	10 – 15 mls/min
Neonates	Purple	26G	<10 mls/min

Cannula`s are used for three purposes

- To provide a route for Fluid or Blood
- To provide a route and drugs
- To allow easy access to veins for blood samples

In the scope of this tutorial we are only interested in the first two uses, as its unlikely the care provider will have the necessary equipment and knowledge to perform blood tests.

Any casualty who may require intravenous fluid or medication requires a cannula. Normally Health Care Professionals will insert one either on route to a medical facility or upon arrival.

Examples of casualties that require them are;
Unconscious and dehydrated.
Have suffered a severe amount of blood loss.
Require IV Antibiotics, Pain Killers (*Analgesia*) or other IV Drugs

Equipment Required

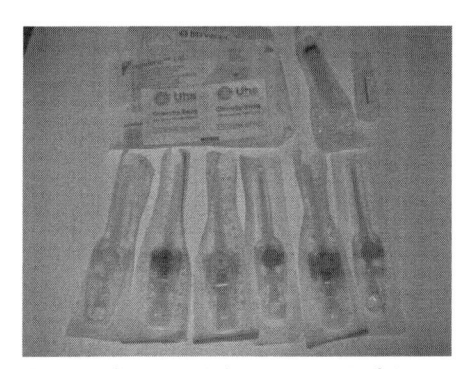

The photo shows a selection of Cannula's, Cannula Dressing, 10ml Syringe, Ampoule 5ml Normal Saline Solution, Sterile Swab. You will also need Latex Gloves and a tourniquet.

Cannulas can be inserted into any vein but the most common sights are in the back of the hand and in the underside of the forearm. The forearm sight offers a larger vein but the elbow may need to be splinted to prevent the casualty from flexing their arm and damaging it.

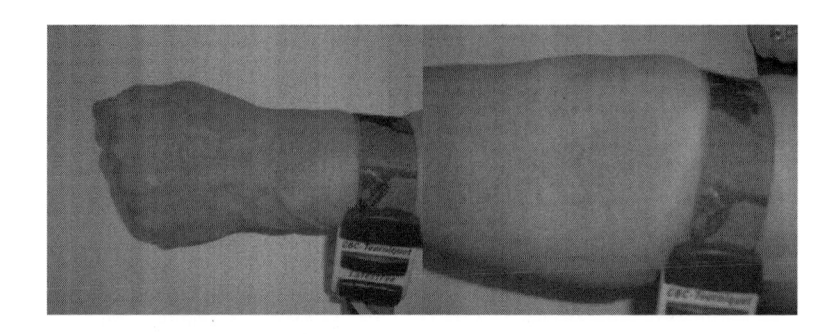

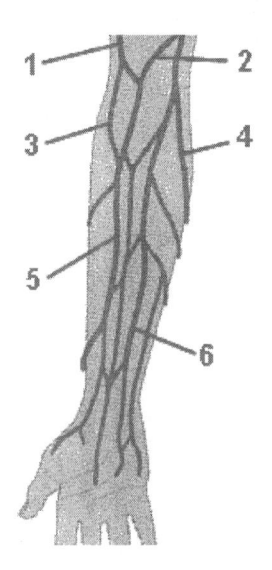

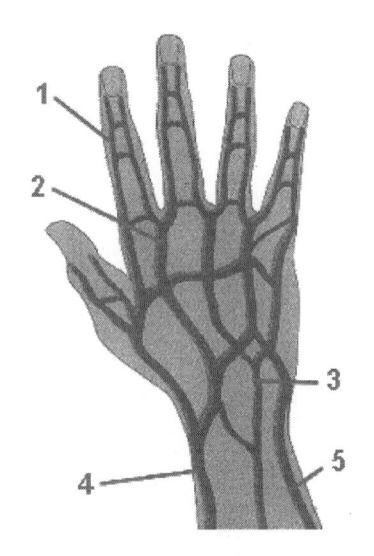

Veins of the Hand

1. Digital Dorsal veins
2. Dorsal Metacarpal veins
3. Dorsal venous network

Veins of the Forearm

1. Cephalic vein
2. Median Cubital vein
3. Accessory Cephalic vein
4. Basilic vein
5. Cephalic vein

4. Cephalic vein   6. Median antebrachial vein
5. Basilic vein

Apply your tourniquet to the casualty's arm about above the elbow, tight enough to stop the return of blood in the veins but not so tight as to impede arterial flow. If they are conscious ask them to clench and release their fist several times, this will cause their veins below the tourniquet to swell (*distend*) allowing easier insertion. Press on the vein with your fingers this if it doesn't move or disappear then you are more likely to succeed with cannulation.

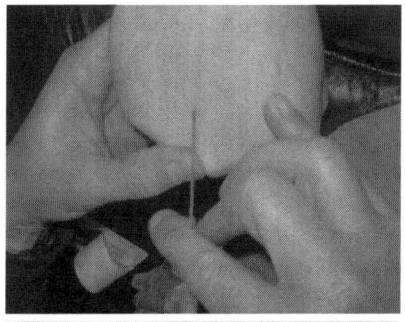

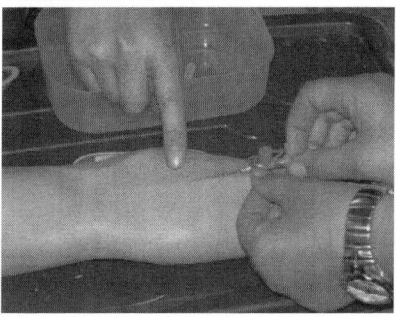

Swab the insertion site with a pre-impregnated swab to cleanse the area and wait for it too dry. Remove the protective sheaf from the cannula and hold with your index and middle finger on the wings with your thumb on the cap, this position should give you good control of the cannula.

With your free hand pull slightly on the skin below the cannula site to prevent slippage, when the skin is punctured. With the cannula at about a 30-degree angle puncture the skin and advance the needle forward into the vein. Once you enter the vein you will see blood enter the chamber at the rear of the cannula. Slowly move the cannula slightly forward being careful not to puncture the other side of the vein. Once you are satisfied you are in the vein remove the cap from the end of the needle and advance the cannula into the vein keeping the needle stationary. As soon as the needle is free from the cannula place the cap at the end of the cannula to prevent blood spillage. Open the top port of the cannula and slowly syringe (Flush) saline into the cannula observe for signs off swelling which would indicate the saline is leaking from the a improperly seated cannula into the surrounding tissues. If the cannula is properly seated place a dressing over the top to secure in place and release the tourniquet.

Injections & Infusions

Why do we need to know about injections and infusions, it is understandable why people stay clear of these and when teaching injection techniques it is obvious there is a great psychological barrier to overcome when sticking a needle in another human being.

There are however circumstance that demand the use of inject able rather than oral drugs, there are other routes and certain exceptions to this but these will be covered in a later article.

Circumstances in which Inject able drugs should be used.

- In an emergency situation where a casualty needs a drug quickly, such as adrenalin in anaphylaxis.
- An antidote or antibiotics in meningitis.

- If a casualty is unconscious or has a low level of consciousness then they will not be able to swallow oral drugs.

- If a casualty is unable to digest the medication due to damage to the digestive tract or nervous system.

- If a casualty is vomiting and unable to keep medication down

- If a casualty has acute diarrhoea where medication may be flushed from the body before it is absorbed.

- If a casualty requires a constant supply of a medication over a period of time.

There are three common methods for injections Sub Dermal, Intramuscular and intravenous.

Sub Dermal injections are useful for local Anaesthetics when suturing or cleaning wounds as well as treating some medical conditions. The Intramuscular and intravenous routes can be used for a most types of medication. Useful examples are analgesia (pain relief), Anti-emetic (Stops sickness), Antidotes, Antibiotics and such drugs as Adrenalin, Anti-histamines, Bronchodilators (Opens Airways) etc

Types of injectable Drugs

Drugs usually come as liquid in individual ampoules, where one or more ampoule may be required to deliver the required dose. The other method is as a powder which is reconstituted using sterile water or saline in its original container then drawn into the syringe for administration.

Getting the equipment together

The basic procedure for preparing the injection is the same for all type. You will need the following;

- A Clean container in which to place you equipment.

- A sterile Syringe bigger enough to accommodate the drug to administered preferably in one go.

- A blunt needle to draw up the drug and another one to administer it sized according to the route used.

- Swabs to clean the skin and Equipment

- And your notes & pen to recode the administration of the drug for future reference.

Procedure for drawing up medication

Check that all sterile equipment and medication is correct, sealed and within use by date before drawing up medication.

If using a Powder within an Ampoule or Bottle, tap the neck gently. To ensure that any powder lodged here falls to the bottom.

If using an ampoule, cover the neck with a sterile topical swab and snap it open. This is to minimize the risk of contamination and to prevent injury to you. If using a bottle remove the top covering the rubber membrane and swab the top of the bottle.

If using a Powder within an Ampoule or Bottle, Inject the correct amount of liquid slowly into the powder within the ampoule and shake it gently until it dissolves and becomes clear.

Check no glass shards have contaminated the mixture and the medication is clear with no bits.

Withdraw the amount required, tilting the container if necessary.

Check the syringe for bubbles and if present tap gently tap until they disperse.

Press the plunger gently to expel any air that is in the syringe.

Change needles to a new sterile one suitable for the route of administration.

Suggested method of vial reconstitution to avoid environmental exposure.

(a) When reconstituting vial, insert a second needle to allow air to escape when adding diluents for injection.

(b) When shaking the vial to dissolve the powder, push in second needle up to Luer connection and cover with a sterile swab.

(c) To remove reconstituted solution, insert syringe needle and then invert vial. Ensuring that tip of second needle is above fluid, withdraw the solution.

(d) Remove air from syringe without spraying into the atmosphere by injecting air back into vial.

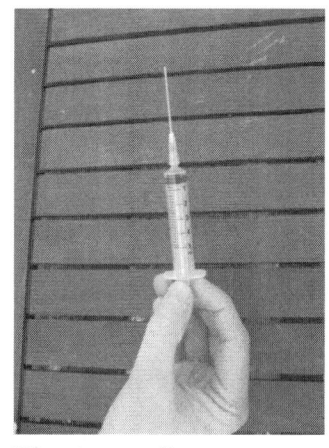

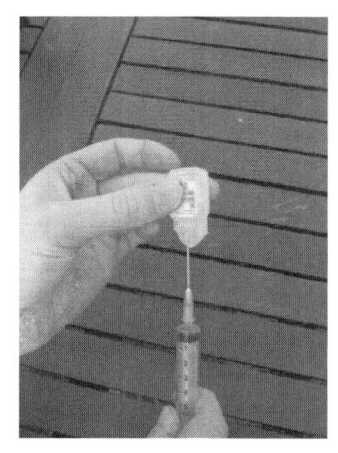

Blunt needles are used to withdraw drugs these prevent shards of glass being scrapped off and accidently drawn up and injected into the patient.

Sub dermal Injections

To perform a sub dermal injection place the patient in a comfortable position and expose the area to be injected

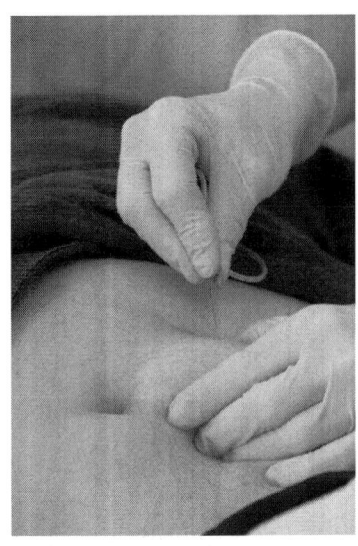

© Anette Romanenko | Dreamstime.com

Clean the site with a swab saturated with isopropyl alcohol 70%. This will reduce the number of germs introduced into the skin by the needle at the time of insertion.

Gently pinch the skin up into a fold. This will lift the subcutaneous tissue away from the underlying muscle to prevent accidental damage and injection into it.

Insert the needle into the skin at angle of 45° (or at 90° for insulin)and release the grasped skin, then inject the drug slowly.

Withdraw the needle rapidly after the drug is administered then apply pressure if there is any bleeding, this will stop blood collection in the tissues.

Dispose of any sharps (Glass ware and needles) into a safe container then record the use, dose and time of the medication

(IF anaesthetizing a wound for cleaning or closure then inject directly into the tissue from the side of the wound. Inserting the needle then releasing the anaesthetic as the needle is withdrawn, if additional cover is needed for a large wound insert the needle again in tissue. previously anaesthetized and advance it through the tissue to the next section of tissue. This process can be repeated around the edge of a wound and the patient should only feel the one needle insertion.)

Intramuscular Injections

To administer an Intramuscular Injection place the patient in a comfortable position and expose the area to be injected

Clean the site with a swab saturated with isopropyl alcohol 70% for 30 seconds then let dry for 30 seconds before injecting. This will reduce the number of germs introduced into the skin by the needle at the time of insertion. It will also reduce the chance of the swabbing solution irritating the tissues or muscle.

Stretch the skin around the chosen site and whilst holding the syringe at an angle of 90°, to the skin quickly plunge the needle into the skin. Imagine you are throwing a dart but don`t let go off it. This will ensure the needle penetrates the muscle.

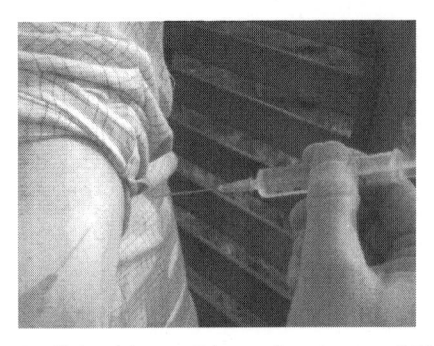

Pull back on the plunger of the syringe. If no blood is drawn into the body of the syringe then depress it approximately 1 ml every 10 seconds and slowly inject the drug. This allows time for the muscle fibres to expand and absorb the medication it further reduces pain and ensure even distribution of the drug.

If blood does appear then you may have hit a vein or another blood vessel, in this case withdraw the needle completely, replace it and begin again.

If the insertion was successful wait 10 seconds before withdrawing the needle. This will To allow the medication to diffuse into the tissue.

When ready to withdraw the needle do so rapidly and apply pressure if there is any bleeding, this will stop blood collection in the tissues.

Dispose of any sharps (Glass ware and needles) into a safe container then record the use, dose and time of the medication

Intravenous Injections

It is neither common practice nor advisable to directly inject into a vein straight from a syringe. The three common methods for administering intravenous drugs are : continuous infusion, intermittent infusion and intravenous bolus injection.

Continuous infusion

Continuous infusion is the intravenous delivery of a medication or fluid at a constant rate over a period of time. This is done to achieve a controlled response. The greater a medicine is diluted the bigger the reduction of irritation to the vein.

Infusions are given by attaching the patients cannula to the fluid via a giving set'.

A continuous infusion may be used when:

The drugs to be administered must be highly diluted such as potassium, which is vital for a healthy metabolism but in a high dose would be lethal.

Maintenance of steady blood levels of the drug is required.

If Fluids are required over a long period

You should only add one drugs to each bag or bottle of fluid and the bag should have a sticky label attached showing what drug and how much has been added. It is important that the drug and fluid are mixed well by inverting the bag several times to prevent the patient receiving a large concentration of the drug in one go. The infusion needs to be monitored and if the solution becomes cloudy stopped at one.

Good care of the cannula site is essential ensuring that it is free of infection, that the cannula remains in the vein and the fluid doesn't leak into the surrounding tissue.

Intermittent infusion

Intermittent infusion is the administration of a small-volume infusion, less than 250ml over a period of generally less than 10 minutes. This dose may need to be repeat to become effective.

An intermittent infusion may be used when:

A peak level of a drug is desires

The drug needs to be at this specific dilution.

The drug become ineffective at a lower Dilution

he patient is restricted in the amount of fluid they are allowed.

All the points considered when preparing for a continuous infusion should be taken into account here, e.g. pre-prepared fluids, single additions of drugs, adequate mixing, labelling and monitoring.

Intravenous Bolus Injection

Intravenous Bolus Injection involves the injection of a drug from a syringe into the injection port of the giving set (a small rubber port) or directly into the top of the cannula. They are injected over a short time period of anything from seconds to 10 minutes.

A direct injection may be used when:

A maximum concentration of the drug is required to vital organs. This is a 'bolus' injection which is given rapidly over seconds, as in an emergency, e.g. adrenaline.

The drug cannot be further diluted or does not require dilution. This is given as a controlled 'stat' injection over a few minutes.

Because of the inherent dangers of giving drugs in this way the manufacturer's instructions should be followed precisely regarding rates of administration (i.e. millilitres or milligrams per minute. If no rate is given inject over 5-10 minutes.

Once again good care of the cannula site is essential ensuring that it is free of infection, that the cannula remains in the vein and the fluid doesn't leak into the surrounding tissue.

Giving Sets

Giving Sets depend entirely on gravity to drive them. The system consists of an administration set containing a drip chamber and utilizing a roller clamp to control the flow, which is usually measured by counting drops. They are by their nature not very accurate and should only be used for giving fluids without medication.

If you are giving medication diluted in fluids you should be using a mechanical delivery system such as a syringe driver or pump to ensure the infusion is given at a precise rate. Unfortunately in a survival situation this option isn't likely to be available.

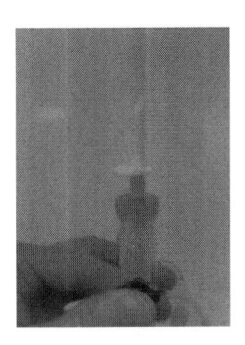

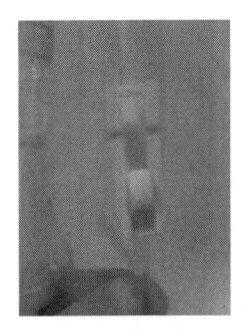

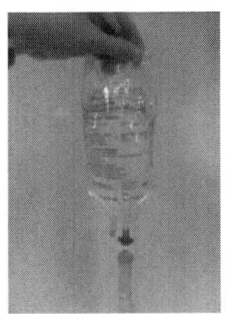

The pressure is dependent on the height of the container above the infusion site as well as the other factors. Roller clamps are used to adjust and to maintain rates of flow on gravity infusions vary considerably in their accuracy.

Flow rate is calculated using a formula which requires the following information: the volume to be infused (how much fluid is in the bag or bottle), the number of hours the infusion is running over and the drop rate of the administration set (which will differ depending on type of set).

The number of drops per millilitre is dependent on the type of administration set used and the infusion fluid. The thicker the fluid the faster the administration/ e.g.

Saline given via a normal 'solution' giving set is delivered at the rate of 20 drops/ml;

Whereas Blood given via a 'blood set' will be calculated at 15 drops/ml.

The rate of administration of a infusion may be calculated using the following formula;

$$\frac{\text{Volume to be infused}}{\text{Time in hours}} \times \frac{\text{Drop rate}}{60 \text{ minutes}} = \text{Drops per minute}$$

In this equation, 60 is a factor for the conversion of the number of hours to the number of minutes.

The big advantages of giving sets are that they are easy to set up, are relatively cheap £3-5 per set and can be safer as its harder for air bubbles to appear spontaneously. They do have disadvantages as they need to be checked to make sure the tubing hasn't kinked stopping the flow of fluid, or flattened by the roller with the same effect. The roller clamp can be unreliable, leading to inconsistent flow rates. If the clamp is left open the fluid will free flow into the patient this can be useful if you have to rapidly give blood or fluid, but is generally not desirable.

Intraosseous

Injecting directly into the bone, used in Cardiac Arrest and in unconscious casualties where IV access is unobtainable. There are several devices available for intraosseous access some examples are; Cooks needles, EZ-IO, Fast IO and Bone Injection Guns (BIGs). They can be inserted into a variety of bones. The Bone Injection Guns are used on the top of the tibia. Locate the tibial tuberosity which is the bump below the knee cap, measure two cm's to the inside of the leg and one cm up towards the knee cap. Remove the red safety cap as shown, place the gun firmly against the leg and pull the 'triggers' to deploy the needle. Remove the cap and attach a three way tap to deliver fluid or drugs, any drugs administered need to be pushed through using a saline flush.

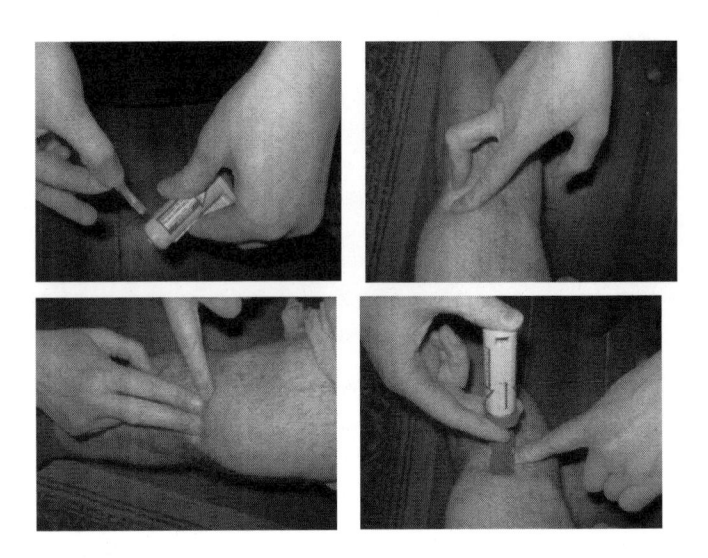

Urinary Catheterisation

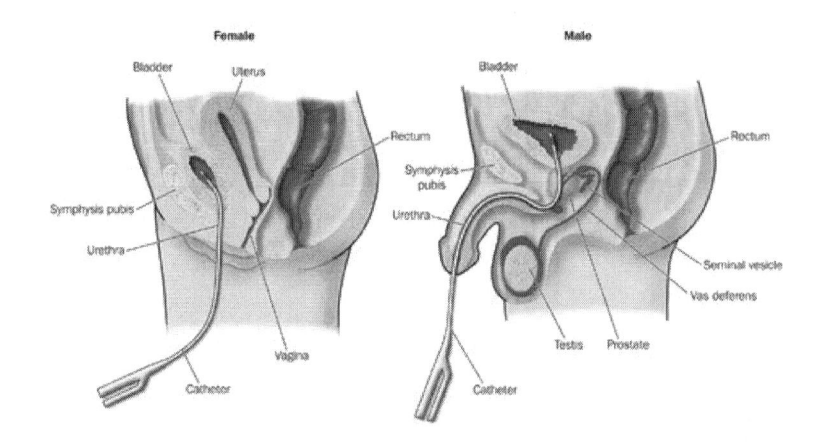

© Legger | Dreamstime.com

There are two main types of catheter:

- intermittent catheter, where it is temporarily inserted into the bladder and removed once the bladder is empty. These have no balloons.

- indwelling catheter, where the catheter remains in place for many weeks.

Catheters are inserted where patients are either unconscious, immobile or have a blockage or problem with their urinary system.

Catheters have double skinned walls, the centre core allows drainage of urine whilst allowing a balloon near the tip to be filled with water to hold it in place within the bladder.

A Catheterization Pack contains sterile gloves, swabs, clinical waste bag, receptacles for urine and cleaning solution.

Resources Required;

- Catheterization Pack or similar
- Selection of Catheters
- Anaesthetic or Lubricating gel

Female Patients.

If conscious ask the patient to lie back draw up their feet then drop their knees to the side. If not try and get patient into a suitable position manually.

Wash your hands and place on sterile gloves, next clean along each side of the labia sweeping from front to back with a swab, use a new swap for each pass to prevent cross infection. Repeat motion cleaning between the clitoris and the vagina. Once area is thoroughly clean discard gloves and put on new pair. An alternative to this is to double glove at the beginning of the procedure and at this point remove the outer pair.

Lubricate the catheter either with lignocaine or a water based gel. Whilst doing this don't handle it if possible, hold it via the plastic inner cover and apply gel directly or using a sterile swab.

In order to pass the catheter you need to find the entrance of the urethra, this is easier to spot in some people than others, but is between the clitoris and the vagina.

The female urethra is only around 4cm long. To advance catheter, hold the distal end over a suitable container until you get a stream of urine. Further advance another 4 cm so tip is well in bladder before inflating balloon.

Male Patients

First wash your hands then place on sterile gloves. With the patient on their back, hold their penis at 90 degrees and pull back foreskin. Clean top of penis and around entrance to urethra use a new swap for each pass to prevent cross infection. Once area is thoroughly clean discard gloves and put on new pair. An alternative to this is to double glove at the beginning of the procedure and at this point remove the outer pair.

Inject the contents of the anaesthetic gel into the urethra and wait for around 3 minutes for it to take effect. Once numb, insert the whole of the catheter holding distal end over a suitable container until you get a stream of urine. Further advance at least another 4 cm so that the tip is well in bladder before inflating balloon.

All Patients

Check the packaging for the correct volume and inject this into the port at the distal end. Then pull gently so the balloon will sit at the entrance to the bladder preventing the catheter from slipping out. Connect the catheter to the urinary drainage bag and hang from bed or place on stand on floor. If the patient is mobile then a leg bag can be used.

Any patient with a catheter should be closely monitored for the presence of a Urinary Tract Infection. Check for cloudy, smelly urine, a raised temperature or signs of confusion.

Nasogastric (NG) Tube

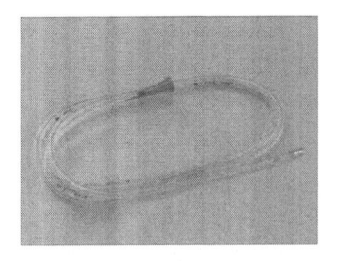

A nasogastric tube or Ryles tube is a plastic tube that is passed via the nose down the oesophagus and into the stomach. It is used either for feeding in a patient who is unable to swallow or for removing the contents of the stomach for instance in cases of poisoning or where there is a problem with the stomach and it is unable to digest food.

Use the smallest bore you can that will do the task required as this will decrease irritation and the risk of aspiration.

- To insert have the patient sit upright.
- Measure the tubing from the nose to the ear, then carry on to the point halfway between the lower end of the breastbone and the navel.
- Lubricate the first 2-4 inches of tube with a lignocaine based gel.
- Insert via Nose, then ask the patient to swallow and advance slowly.
- Stop when recorded length is reached.
- Taking sips of water will help
- If it gets stick turn tube and gently advance
- Stop if patient become distressed

- Check correct position with blue litmus paper, the acid content of the stomach should be sufficient to produce a colour change.

- Secure with tape to prevent slippage

NG Tubes can be kept in place for between 2-6 weeks. They should however be checked on a regular basis at least daily and after prolonged coughing or vomiting bouts.

Feeding should be at a rate of 30mls/hour, special enteral feeds are available which provide all the nutrients the body requires, but rehydration mixtures can also be given via this route.

Recovery Position

The recovery position is designed to keep the casualty in a neutral position, keeping the airway open, allowing vomit and secretions to drain away and keeping chest off floor to aid breathing. To place a casualty in the recovery position, follow the steps below

- Presuming casualty starts on their back.

- Remove glasses and check pockets for bulky objects such as keys that would cause discomfort.

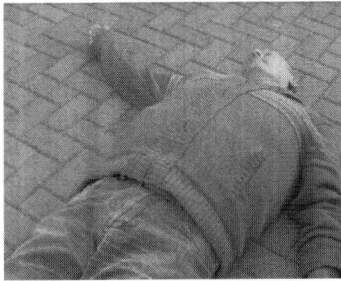

Straighten the casualty's legs then place the arm nearest to you out at right angles to his body, elbow bent with the palm facing up.

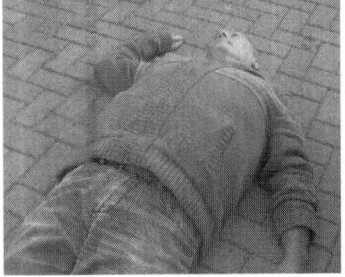

Bring the far arm across the body, whilst holding their hand place it under the cheek nearest to you.

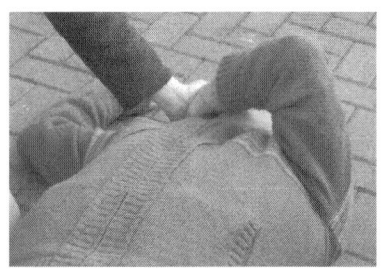

With your other hand, grasp the far leg just above the knee and pull it up, keeping the foot on the ground.

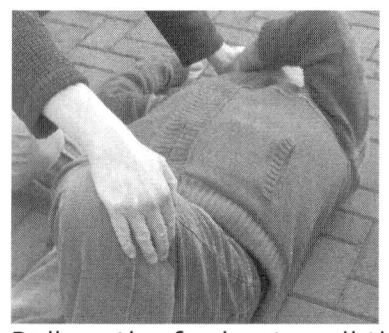

Pull on the far leg to roll the casualty towards you.

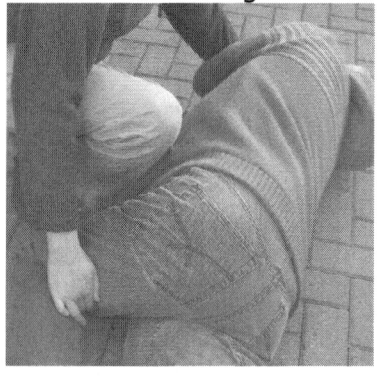

Tilt the head back to maintain a patent airway.

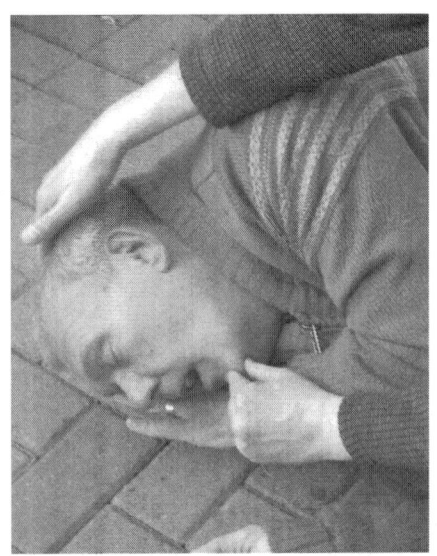

Bend the upper leg so the hip and knee are bent at right angles.

Use back of hand to feel for casualty's breath.

If the casualty remains unconscious for more than 30 minutes roll on other side to prevent pressure areas developing.

Choking

Two forms of choking exist, mild and severe. Ask the casualty if they are choking, if they can answer then the choking is mild; if not and they nod, point to or grasp their neck it is severe.

Mild Choking

In mild choking the casualty feels something in their throat but is still able to cough, speak and most importantly breathe. In most cases just encourage the casualty to cough and this will clear the obstruction.

Severe Choking

In severe choking the casualty feels something in their throat is unable to cough, speak or breathe. They may have an audible wheeze or be unconscious.

If the casualty is unconscious or becomes so during treatment and there is no signs of breathing start CPR see below.

If they are conscious alternate 5 backslaps with 5 abdominal thrusts. Check after each slap or thrust to see if condition has improved, if the casualty is able to breathe stop and encourage them to cough as above.

To administer back slaps;

- Stand behind the casualty.

- Support the casualty's torso with one arm and lean the casualty forwards as far as comfortable

- Give up to 5 blows between the shoulder blades with the heel of your hand.

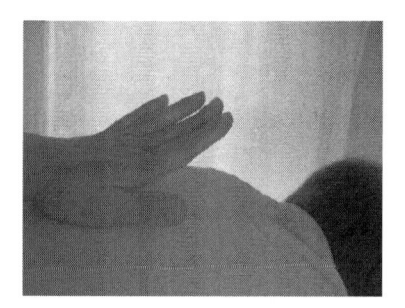

To administer abdominal thrusts;

- Stand behind casualty and put both arms round their abdomen.

- Lean the casualty forwards.

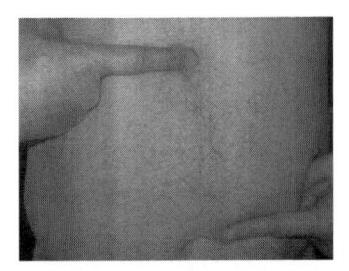

- Clench your fist and place it between the navel and the bottom of the breastbone (sternum).

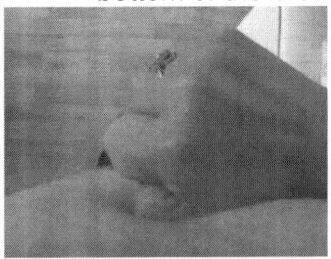

- Grasp this hand with your other hand and pull sharply inwards and upwards with a rolling motion.

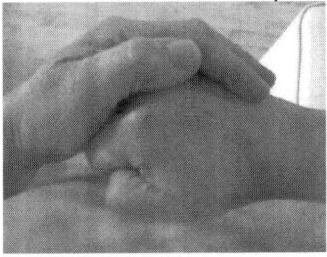

- Repeat up to five times.

Managing an airway

If suction equipment is to manage the airway use this to remove blood and debris from the mouth and throat, being careful not to trigger the gag reflex by inserting the suction catheter too far, which might cause fluid to move down to the lungs, a condition known as aspiration.

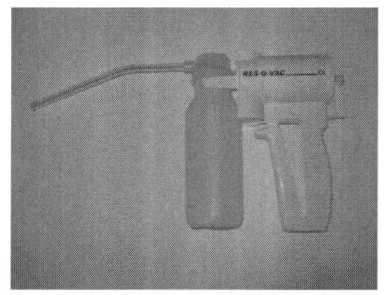

Depending on your level of competence at airway management a number of aids are available.

Oral-pharyngeal Airway (OPAs)

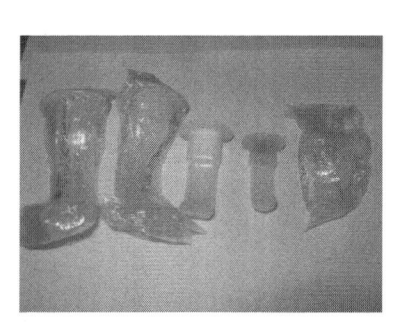

Oral Pharyngeal Airway OPAs are the easiest to insert and require little training or practice. They are slightly curved and flattened plastic tubes that when inserted lay on top of the tongue preventing it falling back and blocking the airway. They are available in a number of sizes and are suitable for infants, children and adults. To determine which size is suitable place one along the line of the jaw the correct size will be the distance between the corner of the mouth and the angle of the jaw.

Sizes and colours can vary with suppliers, common sizes in the UK are;

Colour	Size	Colour	Size
Yellow	5	White	1
Red	4	Black	0
Orange	3	Blue	00
Green	2	Clear	000

To insert an OPA, Use the following aide Memoire Invert, Insert, Rotate, Locate.

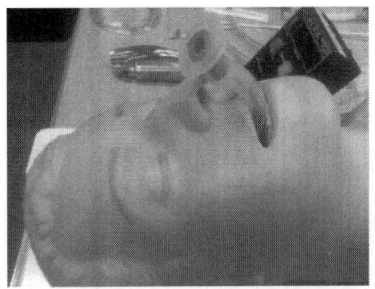

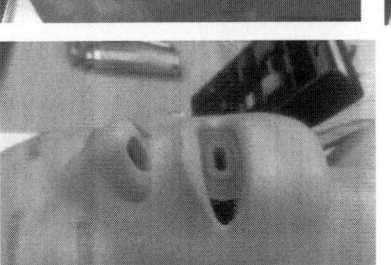

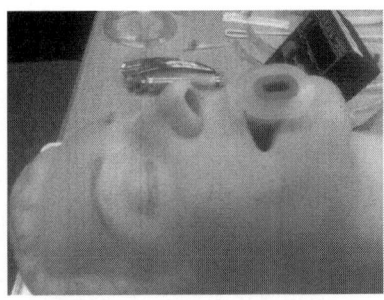

Locating OPA between teeth and lips will keep OPA in position.

Nasopharyngeal Airway (NPAs)

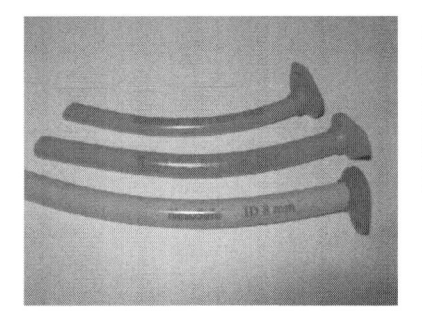

Nasopharyngeal Airways (NPAs) are semi-rigid plastic tubes that are inserted through the nose into the back of the throat.

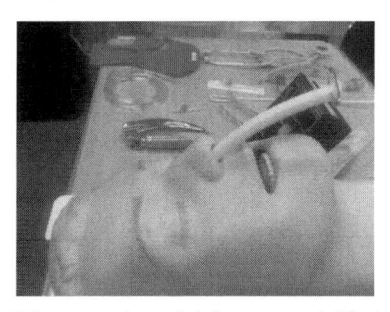

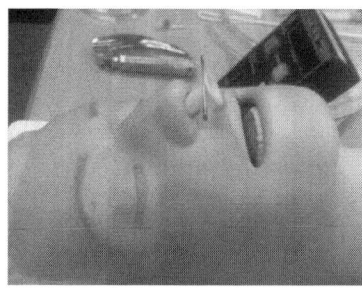

These should be used if there is damage to the jaw or swelling in the mouth which would make insertion of the oral airway difficult. They are either completely soft or with a solid distal end. When inserting the airway, check that the casualty's right nostril is not damaged; then if using the soft variety, place a safety pin through the end of the airway to prevent it being inhaled, insert using a twisting motion. If however the casualty has a fractured nose or you suspect they may have a fracture of the base of the skull, these shouldn't be used as inserting them could cause additional damage. This method is not suitable for children under 6 years of age.

Laryngeal Mask Airways (LMAs)

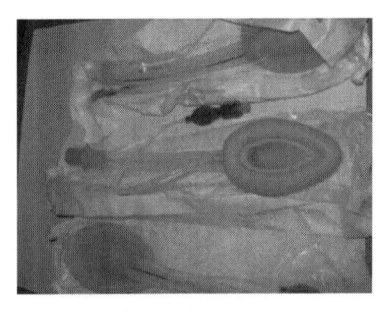

Laryngeal Mask Airways (LMAs) consist of a tube with an inflatable cuff that is inserted through the mouth into the back of the throat (pharynx). They are less likely to trigger the gag reflex and are easier to insert than ET tubes, see below.

Laryngeal mask airways come in a variety of sizes from large adult (size 5) to infant (size 1). There are several different models of LMAs but all work in basically the same way. The end of the LMA has a connector that is compatible with bag, valve and mask and automatic ventilators.

LMAs are widely used in hospitals both for emergency care and some surgical procedures, they are growing in popularity in pre-hospital care too.

Endotracheal intubation

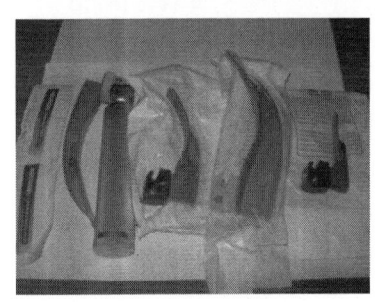

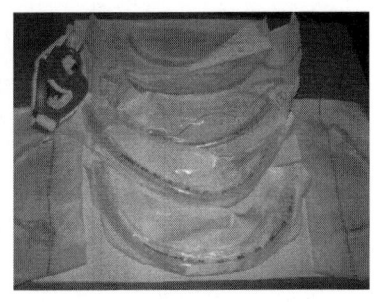

The next method should only be used if you have received training in its use. Endotracheal intubation is where an airway is passed through the mouth and into the windpipe (*trachea*). This is achieved using an instrument called a Laryngoscope, which is placed in the mouth and used to lift the jaw and supplies an illuminated pathway though which to pass the endotracheal airway or *E.T Tube* between the vocal cords.

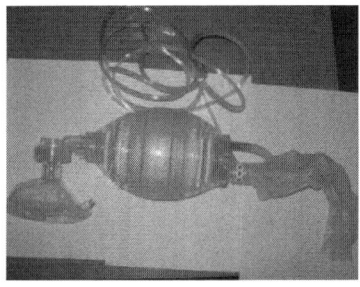

Advantages of this type of airway are that a bag and mask resuscitator may be attached directly to the top of the airway, which allows for easy ventilation of the casualty. If oxygen or a mechanical/electrical ventilator is available this can be also attached. This is the preferred method in hospitals if the casualty has stopped breathing, and can also be used for the direct administration of drugs although the latest resuscitation guidelines is discouraging this practice as injecting directly into the blood stream (IV) or into bone (IO) is preferred.

Cricothyroidotomy

The last method is only rarely use where extensive swelling (*oedema*) or damage Blocks (*Occludes*) the windpipe. It is a surgical method called cricothyroidotomy where an instrument which consists of a hollow tube (*catheter*) over a needle is used to puncture the cricothyroid membrane in the throat and introduce an airway directly into the windpipe. This is a last ditch method as any cut (incision) in the skin bring with it the risks of bleeding (haemorrhage), infection and damage to other internal structures.

Recording a Blood Pressure

The first higher figure (Systolic) is the pressure of blood leaving the heart; the second lower figure (Diastolic) is the pressure of blood in your arteries between heart beats. Blood Pressure is measured as millimetres of mercury, expressed as mmHg

An average normal adult has a blood pressure of 120/70 mmHg. As we get older our blood pressure tends to rise. The treatment threshold at which patients are Hypertensive (have High Blood Pressure) and are usually medicated is 140/85 mmHg. Pressures between 120/70 mmHg and 140/85 mmHg are referred to as pre-hypertensive blood pressures, and above 140/85 mmHg as Hypertensive pressures. People who have had a stroke, heart attack, have coronary Artery disease (CHD) or diabetes should maintain their blood pressure below 130/80 mmHg.

Blood Pressure for Children varies as follows:

New Born	Systolic 70-90	Diastolic 45
6 Months	Systolic 70-90	Diastolic 55
1 Years	Systolic 70-90	Diastolic 60
2-4 Years	Systolic 80-100	Diastolic 60
6-8 Years	Systolic 90-110	Diastolic 60
10-12 Years	Systolic 90-110	Diastolic 65
14 Years	Systolic 100-120	Diastolic 65

Ultimately people die due to decreased cerebral perfusion, which is a decrease in the amount of oxygen going to their brain. This is generally caused by clinical shock which itself has a number of causes:

- Hypovalemia - Loss of Blood, plasma or severe dehydration
- Cardiogenic – Damage to the heart
- Obstructive - Damage to the heart or Lungs
- Septic – Bacterial, viral, fungal infections
- Neurogenic – Due to head or spinal injury
- Anaphylactic – Due to vasodilatation , blood vessels getting bigger

To measure blood pressure you either need a manual or automatic sphygmomanometer. Automatic ones can be affected by a number of things, so best practice is to use a manual one.

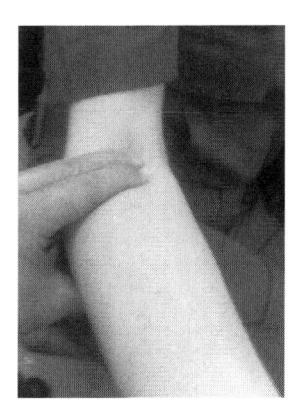

Apply the cuff of the sphygmomanometer to the upper arm

Straighten arm with palm up

Feel for the brachial pulse on the inside of the elbow

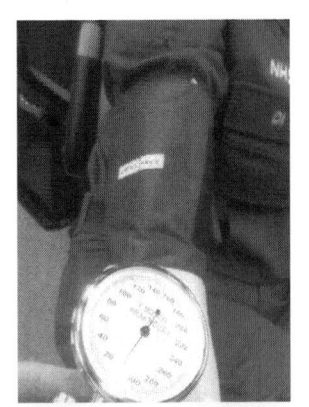

Close the valve

Inflate until you can no longer feel the pulse then increase by 30mmHg

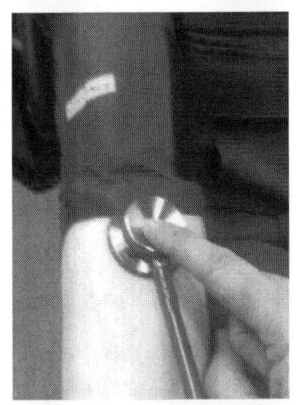

Place the stethoscope over the pulse

Open the valve then slowly release the air.

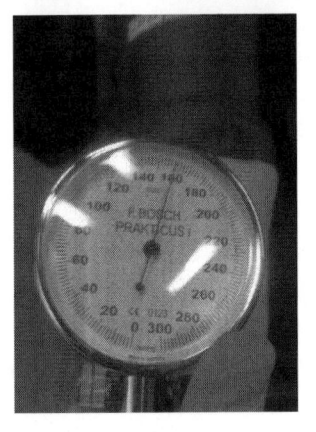

Listen for the sound of blood passing through the vessel (systolic)(thump-thump-thump)

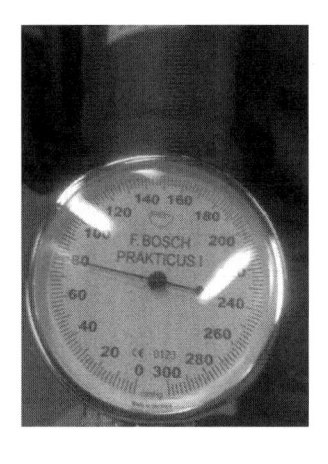

Listen for when it stops.
(Diastolic)

Pulse Oximetry and Oxygen Therapy

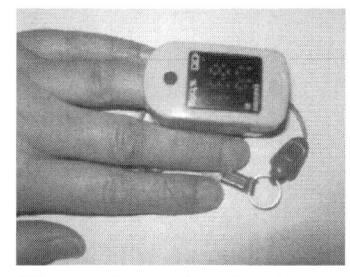

Pulse Oximetry is a way of monitoring the oxygenation of the patient's haemoglobin in their blood. A sensor is placed on a thin part of the patient's body, usually a fingertip, toe or earlobe.

Since the introduction of the British Thoracic Society Guidelines 2008 the amount of oxygen administered to patients has reduced considerably.

Oxygen cylinders are generally black with a white collar. They have one or two connectors, the primary being a nipple and if there is a second it's often a Schrader valve, this allows connection to a ventilator or oxygen piping. On the top is a dial where the flow rate can be altered. Most allow between 1-15 L/min but others are more restrictive. The head also contains a gauge showing how much oxygen is remaining.

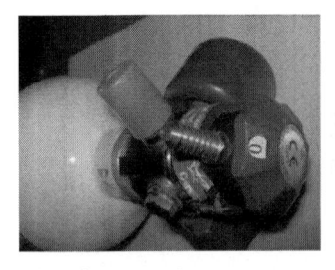

The key point to remember is that oxygen is given to provide adequate perfusion, not to cure breathlessness or push patients oxygen levels above what is normal for them as in the case of COPD.

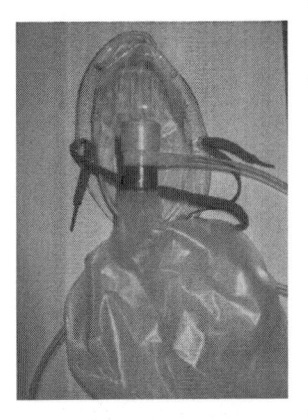

To give maximum concentration of oxygen through a non-rebreather mask first fill the attached bag with oxygen by keeping finger over inlet into mask until bag expands.

Some patients, who are classed as critically ill still need to receive High Flow oxygen i.e. 100% or 15/min through a non-rebreather mask. These include;

- Cardiac Arrest
- Shock
- Sepsis
- Major Trauma
- Anaphylaxis

- Near Drowning
- Carbon Monoxide Poisoning
- Major Head Injury

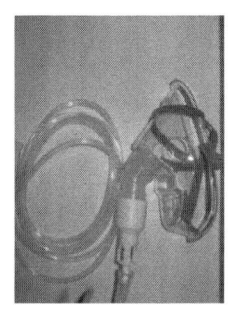

Other than in the above cases if the patient has COPD it likely their body is accustomed to surviving on lower concentration of oxygen. In this case oxygen is given if their oxygen saturation drops below 88%. Oxygen therapy is given via a 28% venture mask at 4L/Min to maintain a saturation of between 88% - 92%

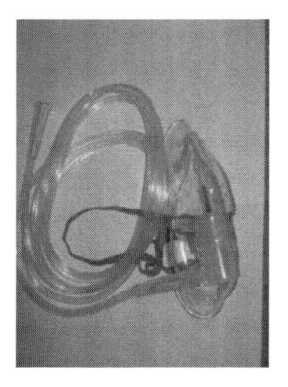

If the patient doesn't have a critical condition or COPD then give oxygen if saturation drops below 94%. Oxygen therapy is given via a simple mask at 8L/Min to maintain a saturation of 94%+. If you are unable to maintain saturations using this method, switch to a non-rebreather mask at 15L/min.

Using a Bag Valve Mask

The Bag Valve Mask (BVM) is positioned over the casualty's mouth and nose. A seal is established by holding the mask with the first finger and thumb and curling the remaining fingers around their jaw. Where continuous ventilation is possible do so at a rate of 10-12 breathes per minute. To achieve the best seal the mask should be held in place with two hand and a second person should squeeze the bag.

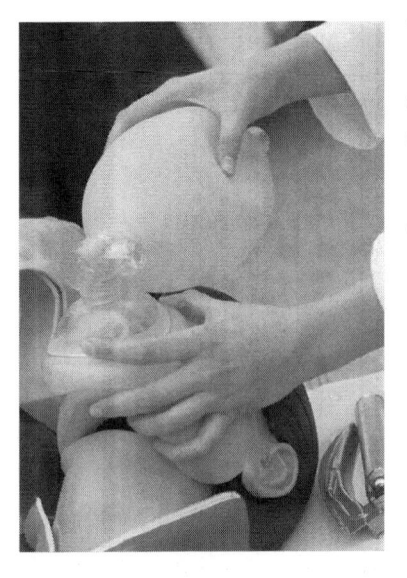

© Mihail Syarov | Dreamstime.com

Defibrillation

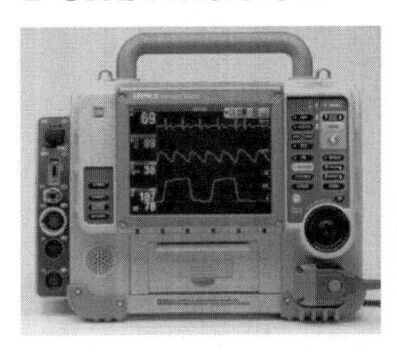

Defibrillation can be achieved using either an automatic external defibrillator (AED) or a manual defibrillator. Some AEDs allow you to monitor the patients ECG rhythm. Manual defibrillators can either monitor and advise on shock or allow manual shocks to be delivered. AEDs generally will not allow manual shocks but some designed for professional use do.

Most modern devices use adhesive pads to deliver the shock as this provides better skin contact, older models may still have paddles, if using paddles apply gel to them before placing on skin.

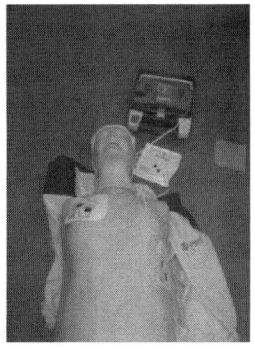

Place pads or paddles, first under right collar bone and the other on the left hand side, where the V6 ECG electrode is normally placed.

If the skin is very wet, dry off. If subject is very hairy and pads will not stick, shave hair. But if a razor is not available do not delay.

If a lump is seen with a surgical scar below the collarbone this may indicate a pacemaker is fitted. Avoid placing the pad or paddle directly over it.

A defibrillator shock stops the heart allowing its natural pacemaker to take over. Multiple shocks may be needed to achieve this.

If using an AED it will determine the required strength of shock required. If using a manual defibrillator you can select the energy setting which is measured in joules.

Two types of defibrillators exist, monophasic and biphasic. Most modern devices are biphasic. The first shock should be between 150-200 Joules with subsequent shocks of between 150-360 Joules. Determining energy setting is based on patient's size and effectiveness of previous shocks. As a general rule start low and work upwards.

There are two shockable rhythms Ventricular Fibrillation and Ventricular Tachycardia both are explained in the ECG section.

During a cardiac arrest the rhythms may change between shockable and non-shockable so must be monitored constantly.

Cardio Pulmonary Resuscitation (CPR)

Information below based on the 2010 Resuscitation Guidelines. This can be divided into Basic, Intermediate and Advanced Life Support (ALS);

General
- Compress centre of chest 5-6 cm and at a rate of 100-120/min.
- Do not interrupt Compressions unless absolutely necessary.
- Give each rescue breath over 1 second
- Check breathing maximum of 10 seconds, normal breathing is 2 good breaths heard in that time.
- Signs of life are coughing, speech or movement

Basic Life Support (BLS)

BLS is CPR using at most a protective device such as a face shield or mask during Rescue Breathes.
- Check casualty Unresponsive
- Shout/Call for Help
- Open the patients airway by placing one hand on their head to tilt it back and two fingers of the other hand, under their chin to lift the jaw.

One of the commonest causes of death is an occluded airway or poor airway management. Remember when using the DR ABC acronym A is the most important.

Without an Airway, Breathing and circulation will soon cease.

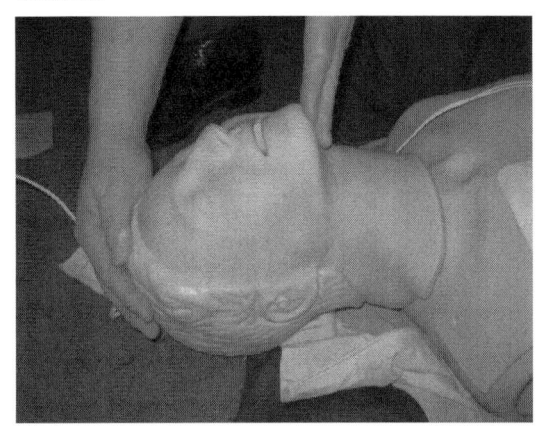

- Check breathing maximum of 10 seconds, by placing your ear close to the casualties mouth and looking down the chest. You will be able to see the chest rise, feel breath on your cheek and may hear sounds of breathing.

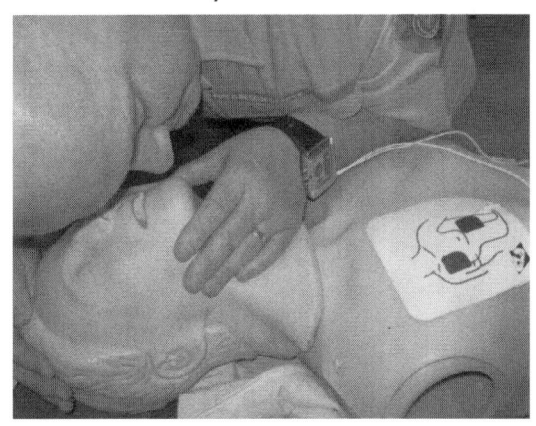

If not breathing Normally
- 30 Chest Compressions
- 2 Rescue Breathes
- Repeat until patient shows signs of life

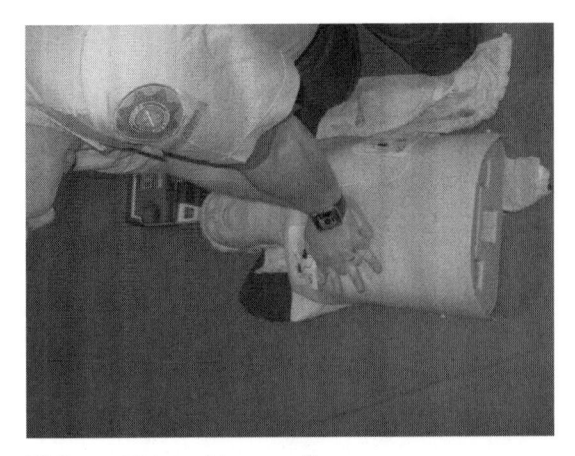

If Breathing Normally
- Place in Recovery Position

Intermediate Life Support (ILS)

As above but using a Defibrillator, basic airway measures and/or Oxygen. There is a blurring of procedures between basic and advanced life support(ALS) dependant on skill and available resources

Advanced Life Support (ALS)

Assess breathing if absent or occasional gasps (agnol) attach defibrillator.

Asses Rhythm;

If Ventricular Fibrillation or Pulseless Ventricular Tachycardia.
- Deliver Shock
- Give CPR 2 Minutes

- Reassess

If Asystole or Pulseless Electrical Activity
- Give CPR 2 Minutes
- Reassess

If Patient shows signs of Life
- Reassess
- Give Oxygen and assist ventilation
- 12 Lead ECG
- Place in Recovery Position
- Monitor
- Treat underlying cause

During CPR
- Give oxygen via bag, valve and mask or face mask
- Consider intubation if skilled in use and it can be done quickly or without interruption of chest compressions (allows continuous compressions.)
- Obtain vascular access either IV or IO
- Correct reversible causes see below
- If still in VF/VT after second shock deliver a third shock then resume chest compressions then give adrenaline1 mg IV and amiodarone 300 mg IV while performing a further 2 min CPR.
- If Asystole or PEA give adrenaline immediately
- Thereafter give adrenaline every 3-5 min

Reversible Causes of Cardiac Arrest

Hypoxia
Give Oxygen and aim for oxygen saturations of between 94-98% once revived.
Hypovolaemia
Give Fluid until Systolic 90mmHg achieved
Hypo / Hyper metabolic imbalance
Give Glucose/Saline to achieve BM between 5-10 mmols
Hypothermia
Warm after the return of signs of life
Thrombosis - coronary or pulmonary
Treat as appropriate
Cardiac Tamponade
Treatment not available pre hospital
Toxins
Give antidote treat as appropriate
Tension pneumothorax
Decompress with needle (see chest trauma)

Precordial thump
A single precordial thump should only be given in the first 5 seconds of a witnessed arrest and must not delay compressions or defibrillation, It is delivered by the bottom of a clenched fist in a sharp blow to the lower half of the sternum from a height of 20cm.

ECG Monitoring and Interpretation

The art or science of interpreting an ECG is a huge subject and many books have been written about it. Some are listed under further reading.

In this section I hope to give you an insight into ECGs, to recognise when one is not normal and what to do about it.

ECG monitoring can be split into three types.
- Single Line of ECG.
- 3 Lead ECG
- 12 Lead ECG

Single line ECGS

Single line ECGS are given by some hand held cardiac monitors and more advanced automatic external defibrillators (AED). They are useful for telling if the heart rate is either too fast or slow, if it is regular or irregular and to have a vague idea about the underlining rhythm. Disadvantages are that they only show what's going on in the heart from one angle and it's very difficult to make any diagnosis from them partly due to the limited view and partly as viewing screen is often quite small. AEDs provide the monitoring capability through the ECG pads that are placed on the patient.

3 Lead ECGS

© Legger | Dreamstime.com

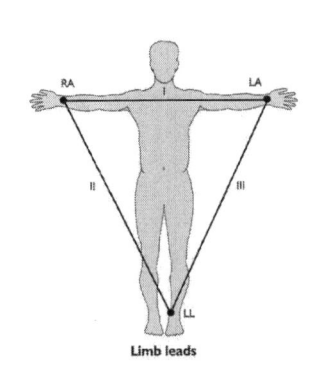

Three lead ECGs are similar as single line ones but show views of the heart at the same time. They may be displayed singularly or one after each other depending on the monitor. The 'Lead' refers to the number of views you get of the heart rather than the number of wires used to obtain it, as 3 Lead ECGs may use three wires as shown or four wires, the last one being an earth lead which usually goes on the right ankle.

These leads are known as limb leads, in the UK they are colour coded as;

- Red on Right Arm
- Yellow on Left Arm
- Green on Left Leg
- Black on Right Leg (optional, but use if present)

The ECG strip is made up of a series of complexes; the closer the complexes are together the faster the heart rate. Widely spaced complexes indicates a slow heart rate. Often a monitor or printed strip will also show the heart rate.

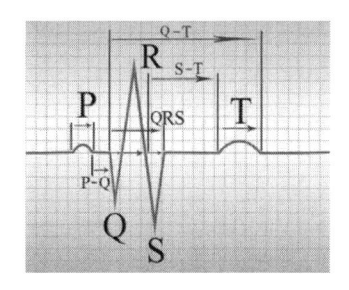

© Mature | Dreamstime.com

The ECG complex is a graphical representation of the electrical activity of the heart. It shows relative positive and negative voltages generated by different areas of the heart as it beats. Although everybody is different there are established normal values. Abnormalities are identified by differences in the shape of the complex.

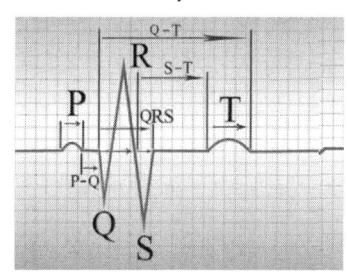

© Mature | Dreamstime.com

Probably the most important ECG change to recognise is ST Elevation. If you compare the two diagrams you will see the base line at point S has increased positively. This indicates the heart is in distress and is the common change present when a patient is having a heart attack. As previously mentioned it is possible to have a heart attack and still have a normal ECG. You can also see the 'T' wave is positive on the first normal ECG and negative on the second. This is also an indication of heart damage and is often seen in angina. It should be noted that ST Elevation is often seen with myocardial bruising and hypovalemic shock not just heart attacks.

12 Lead ECGs

A 12 Lead ECG consist of 10 leads, four limb leads as mentioned above and six chest leads as illustrated below use the standard placement.

© Legger | Dreamstime.com

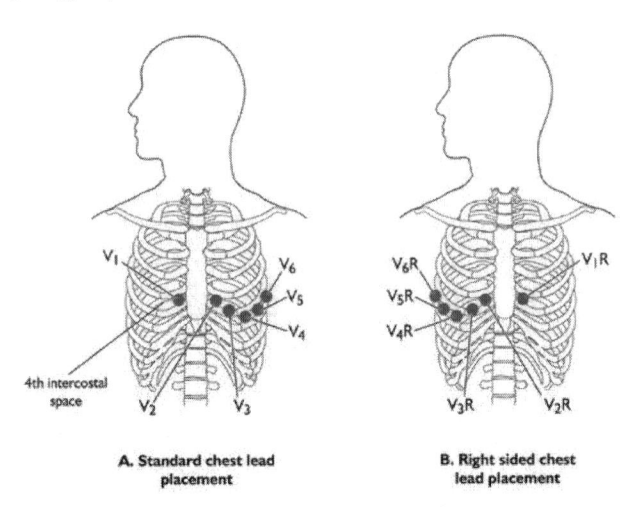

A. Standard chest lead placement

B. Right sided chest lead placement

V_1 Between the 4 & 5 ribs to the right of the breastbone.

V_2 Between the 4 & 5 ribs to the left of the breastbone.

V_3 Between leads V_2 and V_4.

V_4 Between the 5 & 6 ribs in a line from the middle of the collarbone.

V_5 Between leads V_4 and V_4

V_6 On the same level as V_4 and V_5 in a line from the middle of the armpit

Dependant on where the changes occur on the ECG this indicates the area of the heart which is damaged. In the survival setting this is of purely academic interest as it makes no difference to treatment. E.g.

an inferior heart attack would show changes on leads II, III and aVF see table below.

I Lateral	aVR	V1 Septal	V4 Anterior
II Inferior	aVL Lateral	V2 Septal	V5 Lateral
III Inferior	aVF Inferior	V3 Anterior	V6 Lateral

Below is a normal 12 Lead ECG.

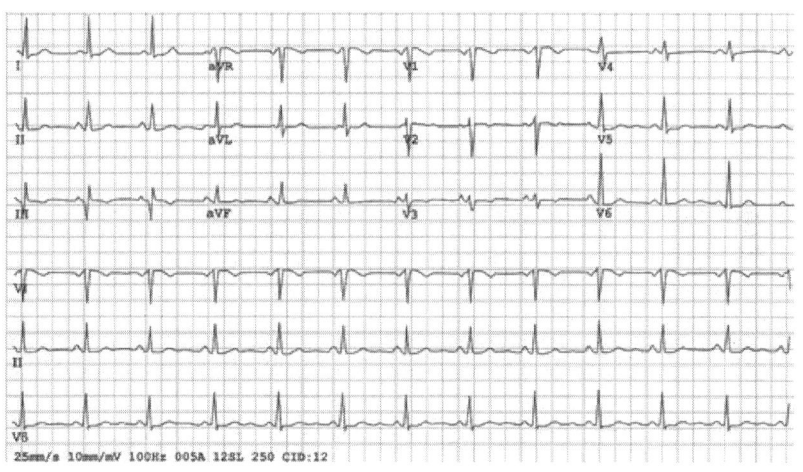

© Inna Ogando | Dreamstime.com

ECG images below reproduced with kind permission from ambulancetechnicianstudy.co.uk

Normal Sinus Rhythm

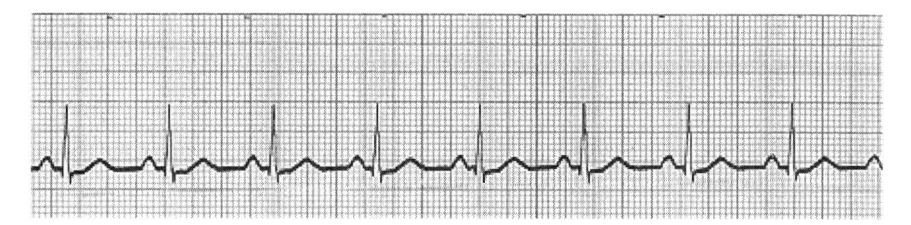

This is the Normal heart's rhythm, the heart can be seen to beat regularly with equal distance between each complex, and each complex is narrow with a rate of between 60-100 beats per minute.

Sinus Bradycardia

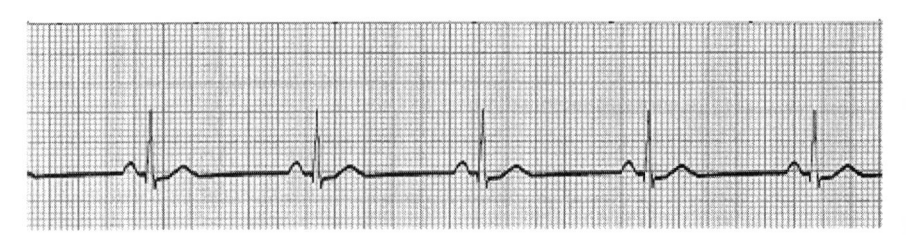

In Sinus Bradycardia, the heart can be seen to beat regularly with equal distance between each complex, and each complex is narrow with a rate of less than 60 beats per minute. Some fit people have a natural heart rate of <60, there is a second level of <40 beats per minute, this is known as absolute Bradycardia. At this level the patient is usually in distress.

Sinus Tachycardia

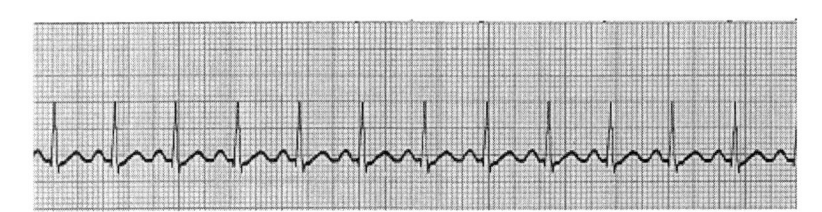

The heart can be seen to beat regularly, with equal distance between each complex, and each complex is narrow with a rate of >100 beats per minute.

Supraventricular Tachycardia (SVT)

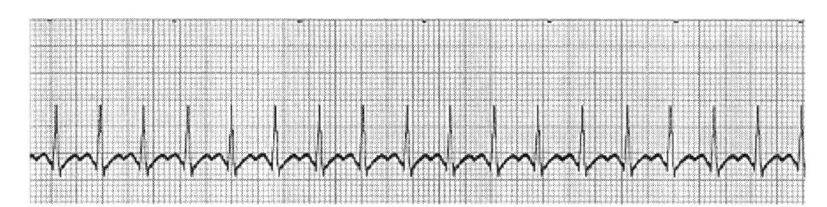

The heart can be seen to beat regularly, with equal distance between each complex, and each complex is narrow with a rate of >140 beats per minute.

Atrial Fibrillation

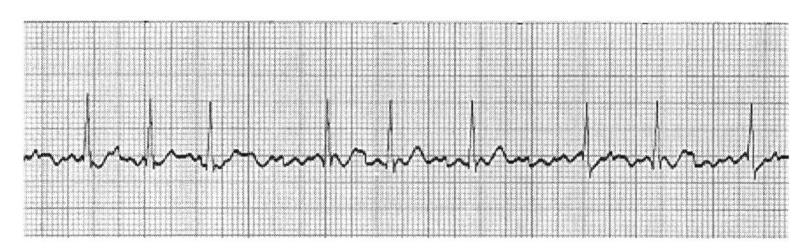

The heart can be seen to beat irregularly, with unequal distance between each complex. P Waves are not distinguishable as the base line is very uneven. Heart rate is usually between 100-140 beat per minute.

ST Elevation

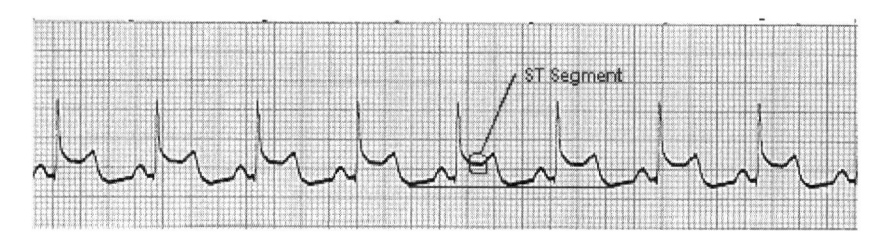

This is the rhythm most associated with a Heart attack, although is not diagnostic unless seen as part of a 12 lead ECG.

Bundle Branch Block

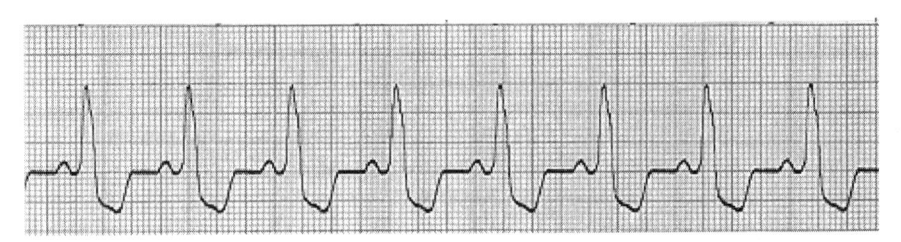

This indicates a problem with the bundles that carry electrical impulse through the heart. This is characterised by a 'rabbit ear' look to the peaks. Right Bundle Branch Block (RBBB) shows as positive changes whereas Left Bundle Branch Block (LBBB) shows as negative changes. A new LBBB may also indicate new damage to the heart.

Ventricular Fibrillation

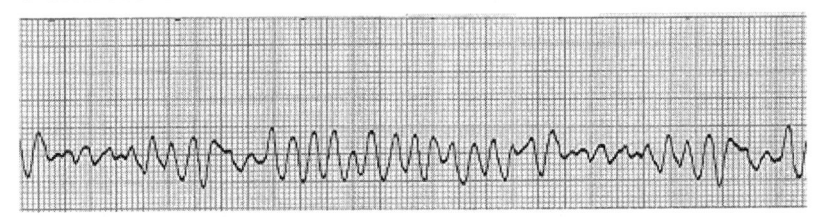

This rhythm is only found in patients in Cardiac Arrest, it shows uncoordinated electrical activity with the heart. These patients need to be immediately defibrillated.

Ventricular Tachycardia

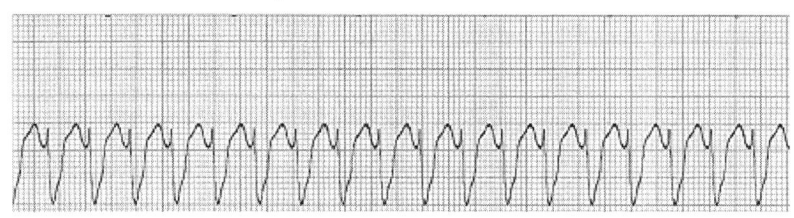

If the patient is in this rhythm and is in Cardiac Arrest. These patients need to be immediately defibrillated. Care must be taken as its possible to have this rhythm and be either conscious or unconscious but not in Cardiac Arrest. The defibrillator only measures electrical activity it doesn't know if the patient has a pulse or not. If it detects a Ventricular Tachycardia of over 180 beats per minute it will consider this as a shockable rhythm. A similar rhythm called Sinus Ventricular Tachycardia (SVT) is shockable if the rate is over 220 beats per minute. In this case the defibrillator may deliver a controlled shock on the R wave of the ECG to stabilise the heart rate.

Managing abnormal Heart Rhythms

Slow Heart Rate (Bradycardia)

An absolute bradycardia is defined as a heart rate of <40 beats per minute. If the patient is unstable i.e. has a Blood Pressure <90mmHg and is losing consciousness treatment should be given. Give oxygen if available and 500mcg Atropine IV every three to five minutes until the maximum dose of 3 mg is given or the heart rate >60 beats per minute. If Atropine fails to stabilise the heart rate the patient will need other drug therapy or Transcutaneous Pacing (TCP).

Fast Heart Rate (Tachycardia)

A fast heart Rate has many causes, shock, Infection, stress etc. If the problem is thought to be primarily from the heart (Cardiac) then there are several treatment options.

Give oxygen, Record an ECG, BP and O2 Saturations.

If Heart Rate >150 and Blood Pressure <90mmHg and is losing consciousness (Patient Unstable) treatment should be given. Give IV Amiodarone 300mg diluted in 10ml water over 10-20 minutes (1ml per one or two minutes) if this fails to stabilise the heart rate there may need cardioversion using a defibrillator.

If Heart Rate >150 but patient is Stable then determine if the Rhythm is broad or narrow complex and if the rate is regular or irregular.

If the complex is narrow and the rate regular encourage Vagal Manoeuvres these stimulate the Vagal nerve and slow the heart rate. Some Vagal Manoeuvres are;

- Blowing down the end of a 10ml syringe to try to raise the plunger.
- Bearing down as if going to the toilet
- Massaging the carotid bodies which are wider areas of the carotid arteries in the neck.

If this fails give adenosine according to protocol starting at 6mg. Adenosine is a chemical cardioverter and may put the patient into cardiac arrest.

If the complex is narrow and the rate irregular it could be atrial fibrillation (AF), consider IV digoxin or IV Amiodarone 300mg diluted in 10ml water over 20-60 minutes.

If the complex is broad and the rate regular give IV Amiodarone 300mg diluted in 10ml water over 20-60 minutes.

Measuring Blood Glucose levels

See diabetes in Medical Conditions. Low blood glucose levels are also present in casualties that are starving, intoxicated or exhausted. Low blood glucose known as Hypoglycaemia is potentially a life threatening condition.

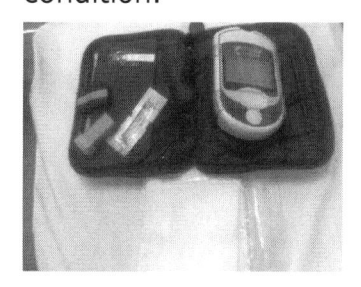

To measure a blood sugar level you will need a Blood Glucose Meter these cost around £30, a test strip and a finger pricker

You need something to prick the finger to obtain a drop of blood. Special safety devices are available but in an emergency a clean, sterile pin or needle will suffice.

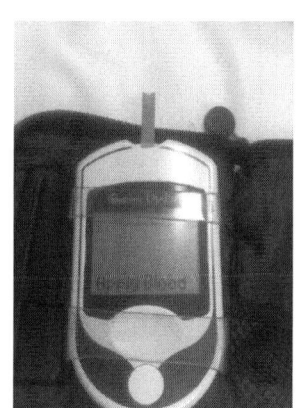

Place test strip in meter the display will say "apply blood" or something similar.

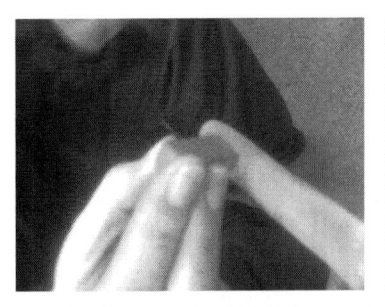

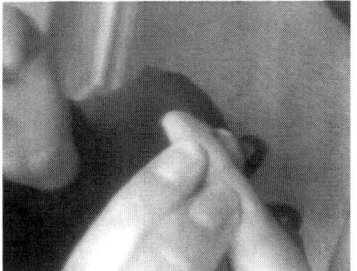

Prick finger, hold finger downwards and apply slight pressure to produce a drop of blood. Stroking the finger with firm pressure will help if blood isn't immediately forthcoming.

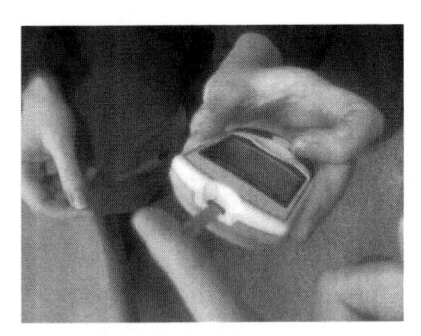

Apply a drop of blood to the top of the strip and wait for the meter to count down, Different models take different amount of time. The model shown takes 20 seconds to analyse the sample.

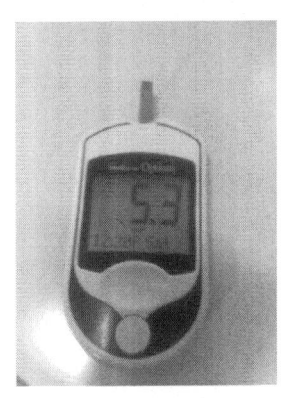

Read and record the blood glucose measurement. If actively treating a patient the level will change quite quickly and can be rechecked ever few minutes.

Chapter 25 Medical Kits & Supplies

Choosing Kit based on perceived risks

When deciding on what Medical Kit to carry, whatever the purpose we prepare for what we consider to be risks. I.e. if your intend to survive in a forest area then an axe and/or saw would be useful and reasonable to carry.

The same applies to First Aid Kit (FAK) & Medical supplies, first compile a list based on the season, activities and likely environment you may encounter. Then look at how lightly a particular situation may occur. If it's very unlikely can you afford the weight / space the supplies that would take to fix it, on the other hand if it's very common can you afford not to carry those supplies to fix it. That must be weighed against its likely hood to severely incapacitate or kill you too of course.

There are also a lot of items that are nice to have but may take up space that would be put to a better use such as high energy rations or trapping/fishing/ water acquisition equipment. Another consideration is can the item be used for more than one purpose thus doubling its worth.

As far as kit distribution goes it may be worth keeping some items in a FAK if space permits which has a use in first aid but also can be seen as a survival item. Then if you lose your survival kit or main pack but have a FAK on your belt or in a pocket then you have those additional resources. Such items could include a Space Blanket, Paper and Pencil, Water purification tablets, small scissors etc

It also excludes medical condition, where you should always carry extra prescription medication. I've also excluded infectious diseases you can only usually catch abroad and such conditions as altitude sickness.

Below we have a brainstormed list of possible risks in no particular order, after each I've put a number as to how likely I think they are to going to occur in a survival situation this is just a personal guess, please make up your own lists and numbers (1 very likely – 5 Very unlikely) based on a temperate climate

Small cuts and grazes (1)

Large Cuts with significant blood loss (4)

Small Burns (3)

Severe Burns (5)

Soft tissue Injury (Strain, Sprain, twist) (3-4) dependant on terrain and activity

Fracture (4-5) dependant on terrain and activity

Minor Insect bite or sting (1) if in suitable climate

Anaphylactic reactions (5) unless very prone to them

Venomous Bite (5)

Other Animal Bites (5)

Heat Exhaustion (4-5)

Hypothermia (3-4)

Chilblains (4)

Frost Bite (4-5)

D & V or Severe indigestion (3-5)

Gum Abscess & Dental Emergencies (5)

'Colds', Chest / Upper Respiratory Tract Infections (URTI) (3-5)

Wound Infections (5)

Headache (3)

Fever (4)

Make your own lists and ranking, depending on type of activity and climate you are planning for. Looking at my list those items which have a ranking of 3 or less are Small cuts and grazes, Small Burns, Soft tissue Injury, Minor Insect bite or sting, Hypothermia, D & V, Colds or Severe indigestion and Headache.

So what do we need to treat these;

Necessary

Plasters; Worth having a few in any kit would go for larger or shaped finger tip ones. As they tend to stay on longer. Most wounds that are smaller enough to be covered by a plaster can usually be stopped bleeding by a small amount of direct pressure and have a low infection risk.

Wound Wipes; For small kits include a few for cleaning when water not available or rationed. Some are alcohol based, therefore flammable and can be used in emergency fire lighting.

Gauze

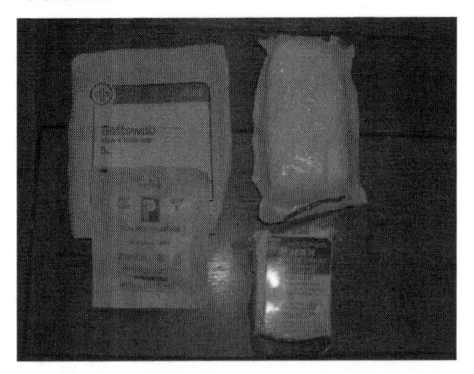

Best to get sterile pieces, packed in 5`s, Can be used for wound cleaning, as dressings, small squares as plaster substitutes, also good as try tinder can be made into charcloth.

Medical Tape; Can be used to secure dressings, as plaster substitute with a piece of gauze, to immobilise fractured fingers or for other cordage uses.

Crepe Bandage; Good for supporting ankles and risks and also be used for securing dressings and tinder.

Triangle Bandages; Useful for supporting limbs, securing dressings. Sunshade etc

Space Blanket or emergency Sleeping Bag to prevent hypothermia or help prevent casualty deterioration.

Pre-Packed Dressings

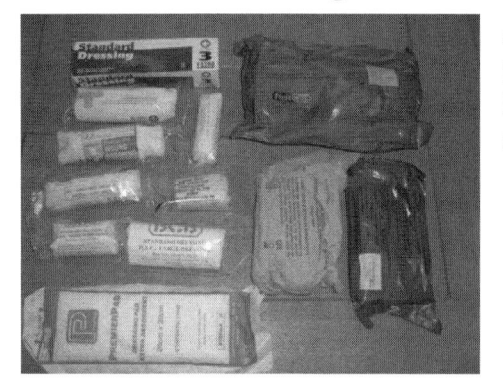

A wide variety of pre-packed Dressings are available

D & V Is very unpleasant and can be a killer in a survival situation, although in normal circumstance its best to let it run its course in survival situations it Can be much more serious. To counter include the following;

Loperamide Anti-Diarrhoea
Prochlorperazin anti-sickness (Anti-emetic)
Rehydration Sachets to replace salts/electrolytes.

Paracetamol or co-codamol Is a useful painkiller for a variety of problems it also reduces temperature in fevers. Don't just stick to one type of pain killer as different drugs work in different ways and having more than one available acts as a balanced analgesia. **Ibuprofen** is a good second choice get the 400mg tabs if you can. Also include a few tablets of a stronger drug such as **Tramadol** if you can. A few **Aspirins** 300mg should also be included.

Nice but no necessary

Butterfly Suture or Steri Strips

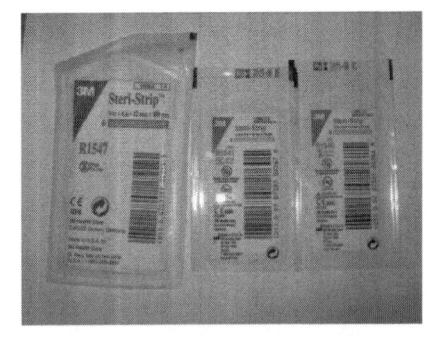

These are used for closing wounds to enhance healing and reduce scar, wounds closed still need to be covered. (only include a few in smaller kits)

Burns Dressings

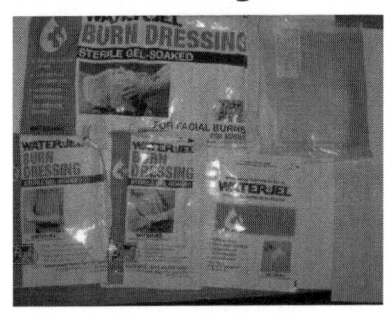

Although they have a cooling and analgesic effect would only include them in larger kits as they are bulky and only work for a limited time.

Sting Relief; nice to have in areas where stinging insects are common

Other medication;

Lemsip, Throat Lozenges, Ranitidine or other Anti-Indigestion Tablets

Other medical supplies should be included based on your risk assessment of potential hazards and the training of available personnel.

Commercial Kits

There are a number of speciality wilderness kits available, most contain a relevant selection of supplies in a nice pouch. They have there place if you want an off the shelf solution however they tend to be expensive and don't contain anything you can't obtain elsewhere. The exceptions to this are Maternity Packs (see below) and Dental kits, a number are described in the section of dentistry.

For a basic kit I would recommend getting a 50 person HSE Refill pack (Which contains a selection of items) such as the one below.

- 1 x HSE Guidance Leaflet
- 60 x Assorted waterproof plasters
- 6 x Standard Dressings (small – eye pads)
- 12 x Standard Dressings (Medium)
- 4 x Standard Dressings (Large)
- 8 x Triangular Bandages
- 20 x Alcohol Free wipes

- 12 x Assorted Safety pins
- 3 x pair of Disposable gloves

Then Top it up with a box of 100+ Assorted Plasters, Box of Antiseptic Wipes and a box of Nitrile Gloves (these are the items you are going to use most off) The 4 items can be bought for less than £20. Then add a selection of Over the counter Medicines.

There is much debate about the value of investing in Resuscitation Equipment for a Preparedness situation its is expensive, requires training and may well never be need also unfortunately the survival rate for pre-hospital Cardiac Arrest is low even in an urban setting.

Entonox (£500)

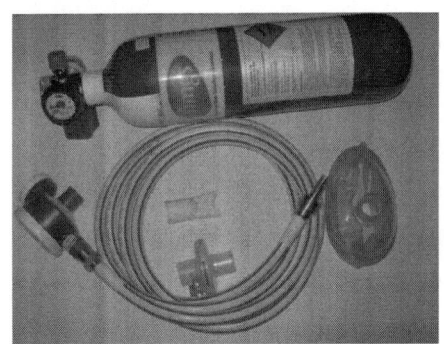

An Entonox set consists of a bag, Gas Cylinder, Tubing with Demand valve and Accessories. It can be self administered either using a mask or a bite piece, either option should include an inline filter.

Oxygen and Airway Equipment

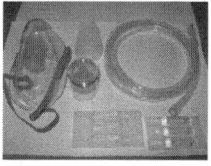

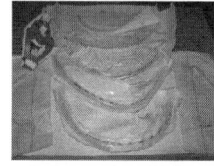

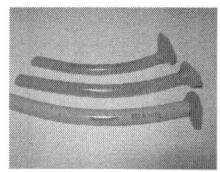

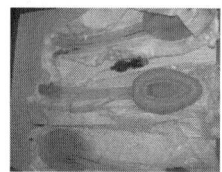

Barrel Bag (£30-50)

Oxygen Cylinder (£130-£150)

Oxygen masks Adult & Childs (£2ea)

Bag and masks Adult & Childs (£8-£40+)

Oral Airways Set (£3)

Nasal Airways (£3ea)

LMAs (£7ea)

ET Tubes (£3ea)

Suction device (£40-£150)

Nebuliser Masks (£2ea)

Administering Medication

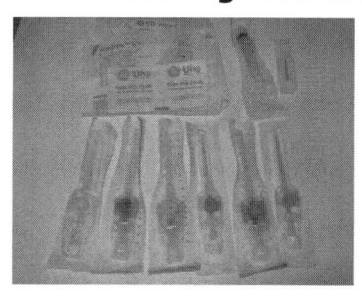

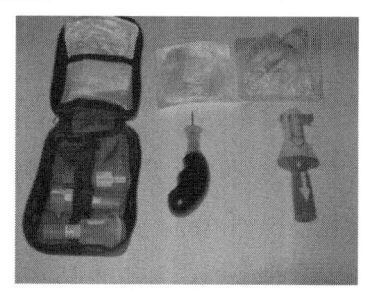

Cannula`s (£1-£2)

Intraosseous Devices £40+

Maternity Pack (£10-£15)

A maternity pack contains the supplies needed for a single uncomplicated birth.

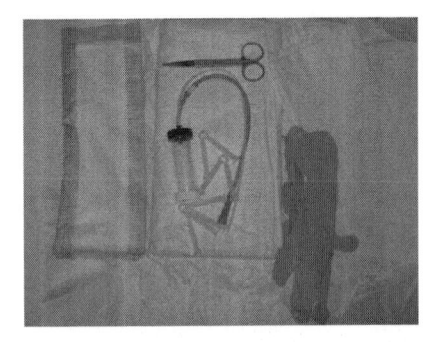

Hand Towel, sterile gloves, Babywrap, mucous extractor, umbilical cord scissors, 4 umbilical cord clamps, maternity pad, Waste bag and Apron.

ENT Diagnostic Set (£25-£50)

1 Light weight C size Handle

1 Otoscope Head double lenses

3 Reusable Ear Speculums
1 Ophtalmoscope Head
1 Bent Arm Illuminator

1 Nasal Speculum
1 Tongue Depressor / Blade
1 Tongue Depressor Blade Holder
2 Laryngeal mirrors

Automated External Defibrillator
(From £800 - £2000+)

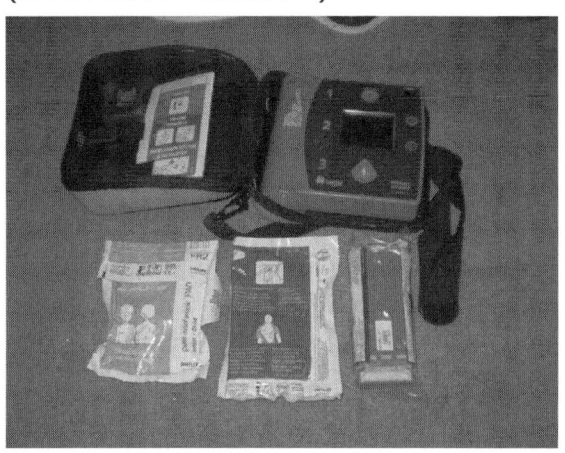

Fracture & Spinal Management

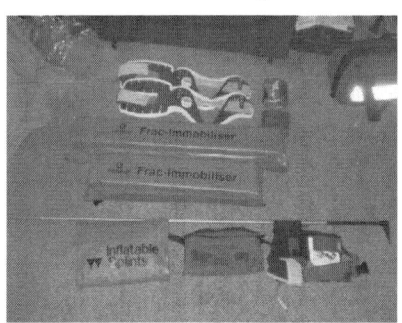

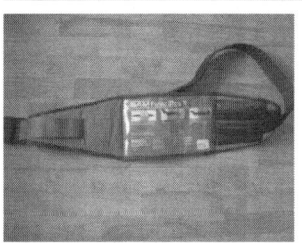

Stiff Neck Collars (£6-£10)

Sam Splints (£10)

Fracture Packs (£30-£40)

Inflatable Splint set (£50-£60)

Kendrick Traction Device (£120)

Sam II Pelvic Splint £50

Wound Closure

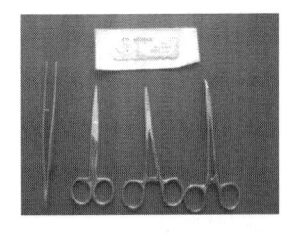

Suture Set (Suture, scissors, needle holder and forceps.) (£7-8) Removal Sets are £2 each

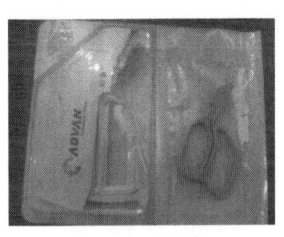

Skin Stapler and removal tool (£8-20)

Diagnostic Equipment

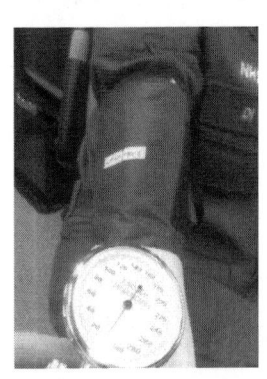

Blood Pressure Cuff (£5-£30+)
Auto Blood Pressure (£15+)
Stethoscope (£5 – £100+)

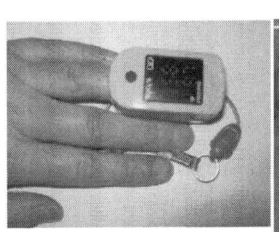

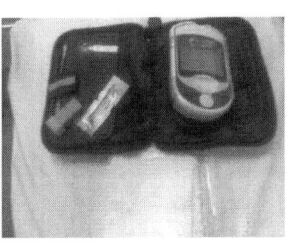

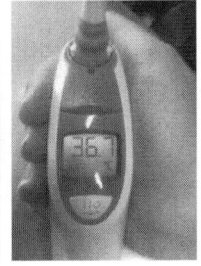

Blood Glucose Monitors are £6-£10 and come with a some test stripes, extra ones can be purchased separately.

Pulse Oximeter (£50-£300) Expensive but a useful piece of compact diagnostic and monitoring equipment.

Digital Thermometers start at £10, cheaper ones tend to be much less reliable, better to go for a brand such as Braun at around £30.

Diagnostic Kits

Without the benefits of laboratory Blood tests to ascertain abnormalities of health, a number of test kits are available which are portable and relatively straightforward to use.

Blood Group

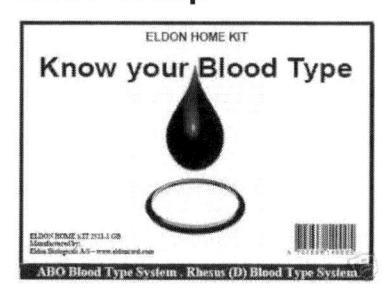

The Eldon Home Kit (£4-5) will determine if your blood group is in the A,B,O,AB and if Negative or Positive.

Cholesterol Level

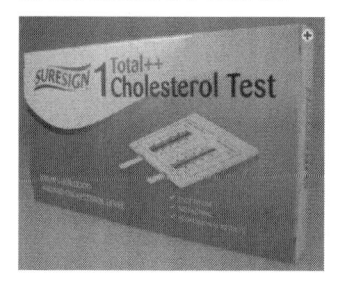

A number of kits are available (£4-10) to test your cholesterol levels. Some only measure your total cholesterol whilst others measure both your good (HDL) and bad (LDL) cholesterol.

Urine Dip Sticks

The strips are designed to analyse a number of different values either 1,2,5,8,10 or 12 tests can be performed from a single strip making them a very useful aid. Each stick has a row of reagent pads which change colour according to the amount of the substance present. Prices vary according to suppliers but you can easily get 100 x 10 Test Strips for under £10 a set.

Tests typically include Blood, Billirubin, Urobiligen, Ketone, Protein, Nitrite, Glucose, PH, Specific Gravity and Leucocytes.

These are not conclusive tests but aid in the diagnosis of Urinary Infections, Kidney Disease, Liver Function, Diabetes and traumatic injuries.

Specialist Diagnostic Kits

Other kits are available to detect;

- Helicobacter pylori which can cause gastritis and stomach ulcers.
- Bowel Cancer through detection of blood in Faeces
- Pregnancy, Menopause, Female Fertility and Sperm Count.
- Illicit Drugs and Alcohol
- Prostate Specific Antigen (PSA) Test
- AIDS/HIV
- Malaria, Legionnaires' disease, filariasis
- Pneumonia and influenza

Wilderness Medical Kit

An example of a more comprehensive kit is detailed below based around a military medics pouch. This is a Day to Day kit designed to Diagnose and treat a variety of Injuries and Illnesses so holds a very limited amount of each item. It should be used as a mobile kit and restocked from supplies at a base camp.

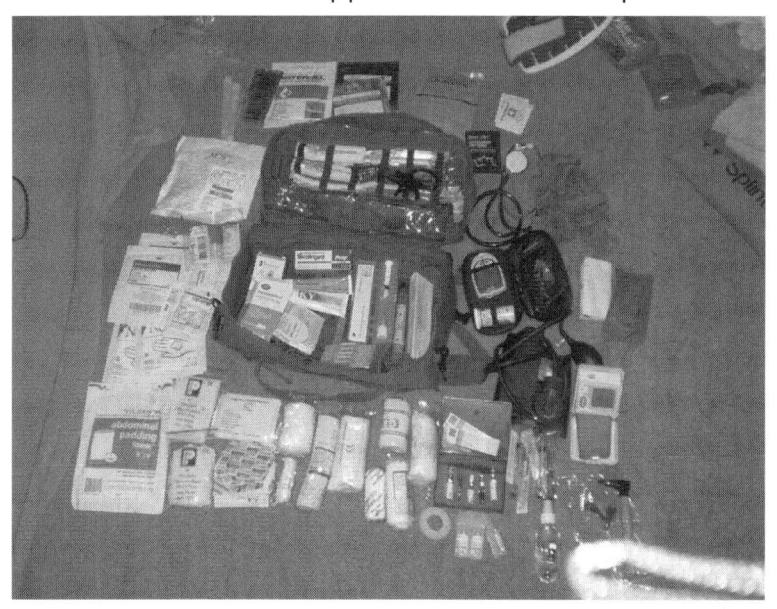

Contents
Cannula's, Adult oral Airways, Nasopharangeal Airways, Syringes, needles, Scalpel, Digital & Mercury Thermometer, Tuff Cut scissors, dressing Scissors, Needle Holder, Forceps, 2 x Saline Irrigation, 2 x Eye Wash

(3 internal mesh bags)

Bag 1 (Drugs)
Glucagen Hypo Kit, Hypstop, Epipen, Salbutamol Inhaler, Paracetamol 500mg, Aspirin 300mg, GTN Spray, Antihistamine Cream,, KY Jelly.

Fluorets, Ameothcaine.

Injectables (Tramadol, Adrenaline, Stemetil, Lignocaine, chlorphenamine)

Syringes and Needles, flushes

Bag 2 (Bandages)
2 large Dressings, 1 no3 Ambulance Dressing, Eye Pad, Micropore Tape
15cm Light Support Bandage, 7.5cm Cotton Stretch Bandage x2, 7.5cm Conforming

Bag 3 (Diagnosis)
Blood Glucose Meter, Sphygmomanometer, Stethoscope, otoscope

Middle Pouches
Assorted Plasters, Antiseptic Wipes, Wound Closures, 2 x Steri Strips, Quickclot, 10x40cm Waterjel Burns Dressing, Asherman Chest Seal, Abdominal Pad 5x9 inches, Triangular Bandages, 2x 5cm Dressings, 2 x 10cm Dressings, 3 Pks Gauze(5), 3 x Petroleum Gauze, iodine solution, Nitrile gloves, 2 Lightsticks, Alcohol Gel, Resus Shield, Personal Protection Pack, Field Dressing, Tissues, Blood Pressure / Pulse Monitor, Pulse Oximeter, Gloves

Chapter 26 Emergency Dentistry

The best person to perform any dental work is a dentist: even other doctors generally have a limited knowledge of the subject and techniques. The information below should only be used in extreme circumstances.

Equipment

There are a number of commercially available Emergency Dental kits that are primarily designed for expedition use. They retail at between £10-£15 each and contain the basic supplies to fill a tooth and perform an examination.

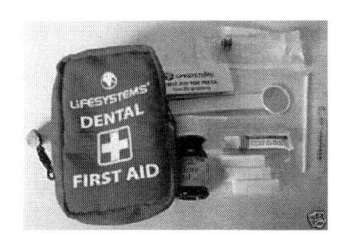

The Lifesystems kit contains: temporary tooth filling compound, syringe and long dental needle, clove oil, mirror, Spatula, Cotton wads and instructions and can be supplied in a hard or soft case.

A similar kit is available called Dr Tooth which allows a greater number of procedures. It contains: temporary dental cement, denture fixative and liner, dental stain remover, temporary filling, Clove Gel, gloves, dental mirror, probe, swabs, and instructions.

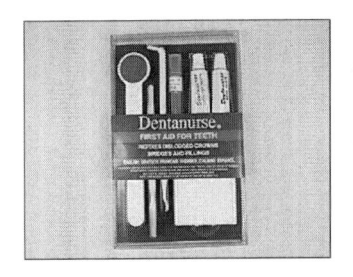

Dentanurse also do a kit with filling material, plastic mirror, mixing tray, spatula, probe, wads and instructions.

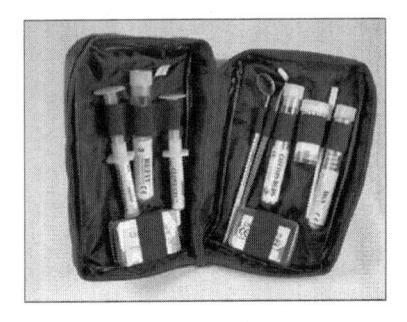

Dr Denti do a nice kit containing dental putty, dental cement, clove oil, mouthwash, + stainless steel dental mirror and probe, gloves, cotton buds, applicator sticks and instructions.

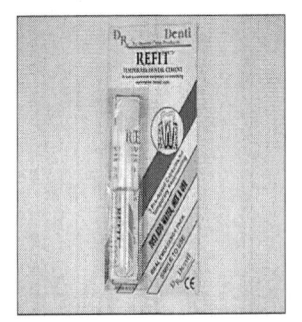

Dr Denti also do a range of individual products to supplement basic kits, such as syringes of filling materials, cement and whitener.

Additional supplies some of which could be used for other medical emergencies include:

Suturing forceps and fine sutures, tweezers, curved surgical scissors, probes, scalpels, set of extraction forceps, elevators or luxators, syringes and 27 or 30 gauge long needles, instrument tray, scaler, curette, dental gags and head torch.

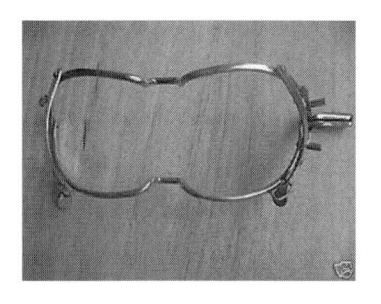

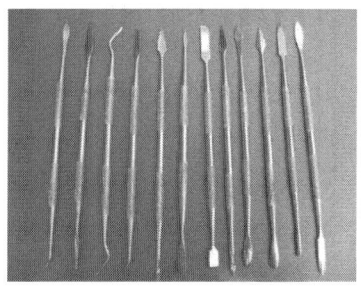

Sterile gloves, gauze to make wads, alcohol swabs, mouthwash, antibiotics, pain killers, local anaesthetic, entenox, clove oil, other filling materials, fluoride powder, toothpaste and brushes, dental floss, etc,

For a large or long term scenario you may consider water jets, suction and dental drills, although using these will be beyond the capability of most people.

You should also have provision to treat respiratory emergencies and anaphylaxis, which will be covered later on.

Drugs and Solutions and Substances used in Dentistry

Most use of drugs in dentistry is for the ultimate aim of reducing pain. Dental pain is caused by swelling placing pressure on nerves, vessels and tissue in the mouth, either in the teeth or the gums. Pain may be activated by heat, cold, or sweet substances and can be intermittent or constant.

Analgesia

The best painkillers for dental problems are NSAIDs such as ibuprofen or aspirin. Ibuprofen can be taken by adults to a maximum dose of 2.4 grams daily i.e. 400mg every 4 hours, although a lower dose may be adequate. Aspirin can be taken by adults to a maximum dose of 4 grams daily, divided down to 4-6 hour intervals. Both drugs should be avoided if you have an allergy or sensitivity to them, particularly if you are also asthmatic.

Alternatives are paracetamol, which can be taken by adults to a maximum dose of 4 grams daily divided down to 4-6 hour intervals of one or two 500mg tablets, or codeine which can be taken by adults to a maximum dose of 240 mgs daily at 4 hour intervals of one or two 30mg tablets. For very severe pain it's possible to take ibuprofen, paracetamol and codeine at the same time as they affect pain in different ways.

Antibiotics

Antibiotics should be used if an infection is suspected and can be used as a prophylactic with a planned procedure on a patient who has cardiac problems, unstable diabetes, renal failure, or who are immuno-compromised.

Choice may be down to availability or known allergies. Metronidazole can be used in conjunction with one of the other four for serious infections.

Antibiotics should be taken for 7 days. Reduction of swelling should occur within 2-3 days. This will decrease pain and doses of analgesia can be reduced to the required level.

Drug	Normal Dose	Severe Infection (Max)
Penicillin	250 mg four times a day	500 mgs three times a day
Amoxicillin	250 mg three times a day	500 mg three times a day
Doxycycline	200 mg initially then 100mg daily	200mg daily
Erythromycin	250 mg four times a day	500 mgs three times a day
Metronidazole	200 mg three times a day	400 mg three times a day

Mouthwashes and gingivitis

When the gum is infected by gingivitis, use a mouthwash such as 'Peroxyl' or 'Corsodyl' for 3 minutes after brushing following a meal. If you haven't got any mouthwash then use a hot salt mixture to swill mouth for 3 minutes. A course of Metronidazole 200 mg three times a day for 5 days is also advisable.

Fungal Infection

The most common oral fungal infection in the UK is candidosis this presents as white yellow plaques on the lining of the oral cavity. If these are disturbed bleeding often results. Candidosis can be treated using Nystatin or Amphotercin in the form of pastilles, lozenges, oral suspensions or ointment. Treatment is given four times a day over 2 to 4 weeks.

Sensitive teeth

To treat sensitive teeth use of a special toothpaste such as 'sensodyne' or mix 1gr of sodium fluoride powder with 50ml of water. Dry the affected tooth, pad either side with cotton rolls then wet tooth with solution for 1 minute. Repeat treatment after once week.

Long term Fluoride treatment

To provide long term Fluoride treatment Mix to a concentration of 2 gr sodium fluoride powder with 1 litre of water. Once a week rinse mouth for 1 minute with teeth closed, washing all surfaces of teeth. Do not swallow or eat for 30 minutes after treatment.

Local anaesthetic [LA] for dental procedures

© Guniita | Dreamstime.com

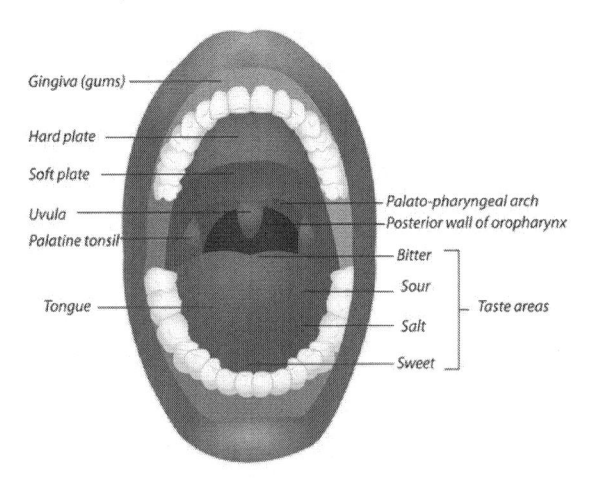

Dental anaesthetic comes in prefilled cartridges. Dental syringes are available from many sources including eBay. Cartridges come in two sizes 1.8 and 2.2 each have dedicated syringes. It is also possible to draw the drugs from ampoules into normal syringes.

There are many different local anaesthetics: septocaine, marcaine and lignocaine to name a few. Each ampoule also contains adrenalin which causes local vasoconstriction. This lessens local bleeding and keeps local anaesthetic near the injection site.

A small gauge needle 27 or 30 gauge should be used to inject.

Anaesthetic can be used for any treatment of painful teeth such as fillings, to give temporary relief from toothache or to perform extraction.

Three common methods of administration exist: Nerve Block, infiltration anaesthetic and intraligamentary anaesthetic.

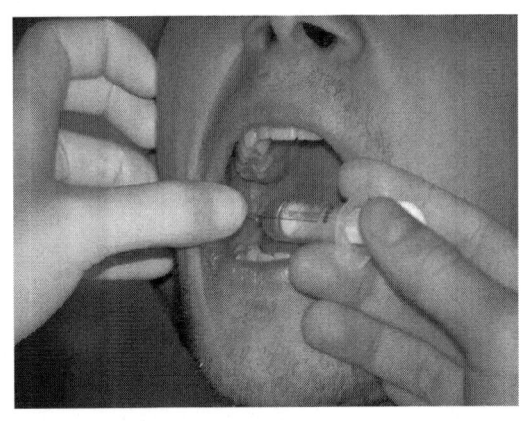

© Zsolt Bota Finna | Dreamstime.com

The most effective block is the inferior block where LA is injected into the muscular pillar at the rear of the molars on the side where work is needed. Insert needle above the back teeth to a depth of 3mms until it touches the mandible. If the needle touches bone earlier or can be inserted much deeper it is unlikely it is in the right place. The whole cartridge should be injected slowly.

The Infiltration methods involve injecting into the gums around the affected tooth both to the front and rear. Inject to a depth of up to 20mm for most teeth and 30mm for canines. Unfortunately this isn't the most effective form of LA and should only be used as an adjunct to nerve blocks.

Cross section of a molar

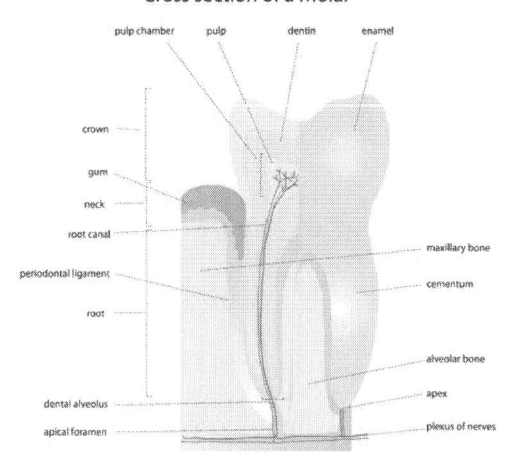

Labels: pulp chamber, pulp, dentin, enamel, crown, gum, neck, root canal, periodontal ligament, root, dental alveolus, apical foramen, maxillary bone, cementum, alveolar bone, apex, plexus of nerves

© Turovsky | Dreamstime.com

When used correctly, injecting LA into the periodontal ligament which surrounds the teeth should provide total LA but can also be used as an adjunct to other forms of LA. Insert a needle 2-3mm into the periodontal ligament and inject 0.3ml of LA. Each of the tooth's roots needs to be injected. Molars have three roots, premolars have two, incisors and canines one.

Each of the techniques can be used in isolation or in combination, and can be repeated until sufficient analgesia is achieved. Seven or eight cartridges is the maximum dose for a fit adult. It may take a while for the analgesia to take effect, up to three hours for a serious infection.

Other forms of conscious sedation can include premedication with a drug such as diazepam, inhalation of gas, intravenous injection and hypnotherapy. Any of these methods should be used as an adjunct to LA.

Varying levels of nitrous oxide mixed with oxygen can cause analgesia and sedation.

15-30% nitrous oxide, Inhalation sedation

25-35% nitrous oxide, Relative analgesia

>55% nitrous oxide, Total analgesia, unconsciousness.

Start with 100% oxygen. Increase nitrous oxide from 10% to 15% at one minute intervals, then from 15% to 20% at a one minute intervals, then increase concentration more slowly. If sensory disturbances occur, reduce concentration by 2.5% to 5%. At the end of the treatment let the patient breath pure oxygen for several minutes. Observe patient for 15-20 minutes to ensure gas has left system.

Dental anaesthetic sets are expensive and difficult to obtain. A viable option could be to use Entonox which is a 50/50 premix of the gases, used as an inhaled analgesia. Entonox sets can be purchased from first aid/ambulance suppliers and Entonox is a pharmacy not a prescription medicine.

Chapter 27 Sexually Transmitted Diseases (STD)

Prevention of Sexually Transmitted Diseases (STDs) is better than cure. It is best to abstain during long trips especially with local people in areas where HIV and syphilis are epidemic. If you do engage in sex always use a condom. Many sexual diseases are symptomatic in males but not in females.

Gonorrhoea

Incubation time 2 to 8 days.

Symptoms in men: greenish-yellow discharge from penis, lowered urine output.

Treatment options

Doxycycline 100mg BD for 7 days or 1g Single Dose. or

Erythromycin 500mg QDS for 7 days.

Also give single dose of Cefixime 400mg or Ciprofloxacin 500mg.

Syphilis

Stage 1: Incubation time 2 to 6 weeks.

Painless ulcerated sores, non-tender enlarged lymph nodes.

Stage 2: Incubation time after 6 weeks.

General skin rash especially on soles of feet and palms of hands, not itchy.

Stage 3: Incubation time after several years.

Multiple Organ Failure.

Treatment

Doxycycline 100mg BD for 15 days.

Chlamydia

Incubation time 7 to 28 days.

Symptoms in men: clear discharge from penis.

Treatment

Doxycycline 100mg BD for 7 days or 1g Single Dose.
Or
Erythromycin 500mg QDS for 7 days.

If diagnosis is uncertain as to STD or UTI give Ciprofloxacin 200mg bd for 7 days.

Herpes simplex

Symptoms: painful ulcers.

Vaginal Discharge/Itching (Monilia)

Symptoms: white discharge with lumps like cottage cheese, itching, burning, increase in amount or foul odour.

Treatment:

Clotrimazole 1% cream, one off dose Diflucan 150mg.

Vaginal Discharge/Itching (Trichomonas)

Symptoms: frothy greenish-yellow, itchy discharge, itching, burning, increase in amount or foul odour.

Treatment:

Metronidazole 250mg TDS for 10 days.

Chapter 28 Reproductive Problems

Testicular Torsion

This is a very painful condition that may be alleviated by twisting outwards. If the pain is due to infection rather than torsion see below. Provide pain and nausea medication.

Testicular infection

This is a painful condition, Provide pain and nausea medication and either Doxycycline 100mg BD or Levaquin 500mg OD.

Obstetrics and Gynaecology

All Obstetric emergencies require urgent evacuation:

Toxic Shock Syndrome

Introduced by tampon into the vagina, usually occurs 2-4 days after period.

Symptoms: fever, circulatory collapse, diarrhoea, rash, thick vaginal discharge.

Treat with respiratory and circulatory support.

Miscarriage

Around 3 in 10 women experience bleeding or cramping within the first 20 weeks of pregnancy, this is best treated with rest. 1 in 10 women will abort the foetus within this period. If the foetus and afterbirth are all passed the pain and bleeding will stop. Check for signs of fever, which may indicate septicaemia - a life threatening condition. Give Rocephin 1g IM followed by 500mg IM BD. If only oral meds are available give Levaquin 500mg OD and pain medication as required.

Ectopic Pregnancy

Two types exist depending on where the rupture occurs either at 6 to 8 weeks or 12 to 16 weeks. A pregnancy test should be performed and will be positive and the patient will have dark blood spotting and cramps, usually diagnosed when lower abdominal pain that occurs suddenly or is intermittent lasts more than 24 hours. Pain can radiate to tip of shoulder, which is often a classic sign. If rupture occurs it causes major blood loss, so early detection and surgery is required.

In the meantime give IV fluids if required to maintain a radial pulse (Approx 90mmHg) and analgesia.

The Pregnant Patient

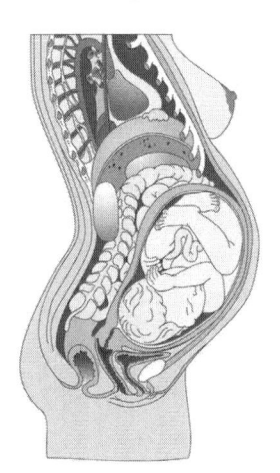

© Maryna Melnyk | Dreamstime.com

The female body goes through many changes when pregnant. Anatomical changes make managing the airway more challenging. They are more prone to gastric reflux both of acid and vomit. Changes in shape of the ribs and physiological changes make them mildly breathless and increase respiration rate. The heart enlarges which can make their ECG appear abnormal and they are often mildly anaemic. Systolic BP is slightly reduced with a larger decrease in diastolic BP.

Due to the size and position of the uterus when the patient lies flat the blood returning to the heart is reduced due to compression of the vena cava. Therefore the patient should be laid leaning to the left (Left Lateral Position) with support under the right buttock.

As they have a larger blood volume for them and the baby, signs of haemorrhage and shock won't start until they have lost a third of their blood volume by which time both will be in serious trouble.

Placenta Praevia

When the placenta implants across the cervical opening.

Symptoms: bleeding after 24 weeks.

Treat in left lateral position, support ABC, NBM.

Placenta Abruption

Occurs when placenta partially separates from the wall of the uterus causing bleeding.

Symptoms: lower abdominal pain, haemorrhage(internal?), uterus tender and rigid, sign of shock, reduced/absent foetal movement or heartbeat.

Treat in left lateral position, support ABC, keep nil by mouth (NBM).

Pre-Eclampsia
A rise in blood pressure due to pregnancy.

Symptoms: Over 20 weeks gestation, diastolic BP>100mmHg or BP>140/90, upper abdominal pain, headache, visual disturbance.

Eclampsia
Fitting in Pregnancy and due to High Blood Pressure.

Symptoms: treat in left lateral position, give o2, 5-10mg Diazemuls

Emergency Child Birth
Normal Labour
First Stage of labour
Cervix begins to dilate, after first 3cm dilation progresses 1cm/hr for single baby and 2cm/hr for multiple births.

'Show' of blood stained mucus.

'Water's break'. Amniotic sac ruptures.

Contractions begin initially 20 minutes apart reducing to 2-3 min intervals.

Baby's head descends into pelvis.

The first stage ends when the cervix is fully dilated about 10cm.

Management
Reassure patient.

Give entenox.

Encourage patient to inhale at beginning of contraction to achieve maximum effect.

Second Stage of labour

The second stage of labour lasts from full dilation until the baby is delivered. Usually lasts between 1-3 hours, usually begins with an urge to push.

The baby's head appears at the vulva. Its head rotates to face towards the mothers back. The back of the head delivers first followed by the face. The head then turns sideward to allow the shoulders to pass.

The shoulder nearest the front is delivered first followed by the other. The rest of the baby is then delivered.

Management

Support the head.

Once head delivered, check cord isn't around neck. If so, slip over the head or double clamp and cut cord in between clamps.

Push gently on the head towards the rear will help deliver the first shoulder. Then push gently towards the mum's front to deliver the other shoulder.

Deliver the rest of the baby.

Wrap the baby and place on mother's stomach or on bed.

Stimulate the baby to cry by tickling feet if not already

Suction airway only if obstructed.

Cut cord after 3 minutes between two clamps.

Control bleeding from tears to the mother with a sterile pad.

Third Stage of Labour

Don't pull on the cord

Placenta 'after birth' will deliver within 20 minutes

Placenta delivery is stimulated by putting baby on mothers breast.

To record the level of the baby's response and to monitor its responsiveness the APGAR score is used.

Sign	0 Points	1 Point	2 Points
(A)ctivity	Absent	Arms and Legs Flexed	Active Movement
(P)ulse	Absent	< 100 bpm	> 100 bpm
(G)rimace	No Response	Grimace	Sneeze, cough, pulls away
(A)ppearance	Blue-gray, pale all over	Normal, except for extremities	Normal
(R)espiration	Absent	Slow, irregular	Good, crying

A score is taken at 1, 5 and 10 minute if required/
A score of:
7-10 is normal,
4-7 may require some resuscitation.
3 or below requires immediate resuscitation.

Retained Placenta

The placenta should be delivered within 10 minutes but if it still hasn't come out after 30 min's it may need some help. If it's loose within the vagina, remove manually. If uterus is contracted, may deliver once cervix relaxes or give Syntochinon 10 units IM.

Where Significant Postpartum haemorrhage where placenta is retained:

- Massage abdomen to encourage uterus to contract
- Put baby to mother's breast.
- Give Syntochinon 10 units IM
- Deliver placenta by pulling gently on cord
- Give IV fluid
- Administer oxygen

Shoulder Dystocia

This is when the shoulder facing forward becomes stuck under the pelvic bone. There are a number of manoeuvres to aid in freeing it. A useful acronym for some of these is HELPER-R:

Help
Call for help if available.

Episiotomy (Delay until other manoeuvre attempted) see below.

Legs

Flexing the mother's legs to her stomach. This widens the pelvis for 30-60 seconds (McRoberts Manoeuvre). If unsuccessful have someone else apply pressure to the lower abdomen (suprapubic pressure), and gently pull the baby's head.

Enter

Position hands in vagina

Two fingers on front shoulder

Two fingers on back shoulder

Rotate anti-clockwise for 30-60 seconds (Rubin Manoeuvre)

Rotate clockwise for 30-60 seconds (Wood-Screw Manoeuvre)

Remove the posterior arm

Roll the mother over onto her hands and knees

Position hands in vagina

Two fingers on front shoulder

Two fingers on back shoulder

Rotate anti-clockwise for 30-60 seconds (Rubin Manoeuvre)

Rotate clockwise for 30-60 seconds (Wood-Screw Manoeuvre)

Replace

If all manoeuvres fail, replace baby's head (Zavanelli Manoeuvre), as an emergency Caesarean section will need to be performed.

When performing any manoeuvre avoid the following:
- Putting pressure on mothers abdomen
- Excessive pulling on babies head or neck
- Twisting or bending babies neck

Episiotomy

This is a surgical cut in the perineum, the muscular area between the vagina and the anus. If the baby is distressed or stuck or if the baby's head is too large in relation to the opening and you think the tissue will tear very badly unless the opening from the vagina is carefully enlarged, then perform a Episiotomy. Either of two common cuts is made either in a straight line from the vulva towards the anus in a 6 o'clock position or preferably between the 7 and 8 o'clock position. Once the baby is delivered and stable, the incision can be closed with sutures.

Prolapsed cord

If the cord is visible through the vulva this is a life-threatening emergency for the baby, as the foetal circulation may be effected. Avoid handling the cord as it may cause it to spasm; place a large wet pad over it to keep it moist. If evacuation is not possible try to place it back inside. Carefully monitor mother and baby until evacuation is possible. Normally an emergency Caesarean section will need to be performed.

Breech Presentations

If another part of the baby presents rather than the head this is known as a breech presentation. You may see baby's: buttocks, genitalia, soles of feet or any limb. If feet are visible lay mother on back with legs flexed towards stomach with the hips and knees flexed and thighs apart. Once nape of neck becomes visible, lift feet gently to aid delivery of head.

If in other breech positions there are three options:
- External rotation of baby
- Internal rotation of baby
- Emergency Caesarean section

External rotation of baby

The hands are placed on the mother's abdomen supporting the baby. The baby is moved up and away from the pelvis and gently turned in several steps from breech, to a sideways position, and finally to a head first presentation. A doctor or midwife usually performs this manoeuvre.

Internal rotation of baby

Same procedure as for external but the practitioner places their hands inside the cervix for the manoeuvre. A doctor or midwife usually performs this manoeuvre.

Chapter 28 Planning for future medical needs

Just like any machine our bodies work better if properly maintained, but every machine has a finite life and sooner or later bits where out. Everybody knows what we should be doing to keep our bodies healthy, good diet, regular exercise, not abusing it too much etc.

But sometime fate gets the better of you as part of your preparations its worth looking at your groups medical history, is there a family history of diabetes, Strokes or Heart Disease, it's possible they will be more prone to developing these diseases yourself. It's estimated that there are currently over 1 million undiagnosed diabetics in the UK at the moment plus many others with risk factors for Cardiovascular disease. I would recommend that all members should get their Blood pressure, Glucose and cholesterol level checked, particularly people over 40 or those with a family history of diabetes, strokes or heart disease.

- Keep Alcohol consumption below recommended levels (21 Units for Men, 14 Units for Women/ week, 3-4 units 1.5 pints of 4% beer, 2-3 units is 175ml glass of 13% wine).
- Replace saturated fats with monounsaturated fats,
- Eat fish twice a week or other source of Omega-3,
  (e.g. mackerel, herring, kipper, pilchard, sardine, salmon, or trout).

- Eat 5-7 Potions of Fruit or vegetables a day
- Reduce Salt
- Eat more wholegrain foods
- Consumption of 2mg of foods a day with plant sterols and stanols
- Participate in 30 minutes of Moderate activity at least 5 times per week
- Don`t Smoke
- Reduce weight if Body Mass index (BMI) is greater than 25KG/m^2, Obesity is over 30KG/m^2 The body mass index (BMI) is calculated by dividing the body weight in kilograms by the square of the height in metres.
- Reduce waist size below 102 cm for Men and 88 cm for Women.

Chapter 29 Long term Care

In a survival or expedition scenario you are unlikely to have the resources to set up a a recuperation area, if a team member needed long term care they could be evacuated as soon as possible.

If you are fortunate enough to have a base camp or are in a position to pre-stock a facility the following should be borne in mind;

The principals of long term care can be split into three sections;

1/ Medical Management of the condition or injury

2/ Symptomatic Relief of ongoing condition

3/ Maintenance of Activities of daily living

Medical Management

The medical management is dependent on the condition or injury, it may be mainly managed with drugs, immobilisation or traction for an injured limb etc

Symptomatic Relief

Again varies with the condition, pain is often a symptom associated with illness this can be managed by a variety of means, see Pain Relief Section. Other symptoms that can be relieved include fever, diarrhoea, nausea, vomiting, shortness of breath, itch etc

Activities of daily living

The Activities of daily living are a Nursing Model of self care Activities which may be decreased if a person's ill or injured. The patient is assisted in those areas in which they cannot fully self-care. This may involve the use of additional resources or modifications to their normal routine.

Maintaining a safe environment (mainly for a patient with reduced mobility)

Remove trip hazards, place safety handles to aid in bathrooms, Use a raised toilet seat, Ensure adequate lighting.

Communication

If the patient is unable to speak due to laryngitis, broken jaw, facial injuries etc, provide a pen and paper. If the patient is short of breath do not pose a question that requires a long answer.

Breathing

If a patent is breathless, conserve energy, if they can sit or lie rather than stand to do an activity let them. Do strenuous activities at start of day when energy is at a peak. Place items frequently used within easy reach to limit bending and stretching.

Diet

Ensure an adequate diet that will promote healing, avoid excessive carbohydrates if exercise levels are reduced. Ensure a good fluid intake. Avoid foods that may cause upset. If chewing is restricted have a soft alternative or puree food.

If patient unable to swallow use a naso-gastric tube (see clinical skills) to feed.

For an unconscious patient use IV fluids to maintain hydration and glucose levels.

Elimination

If mobility is limited patient may not be able to make it to bathroom, so a commode, bedpan or urine bottle should be provided. Disposable products are available but they are designed to be mashed by a special machine so washable ones are a more realistic option. If you need to use them in a vehicle, absorbent granules can be used in the container which absorb moisture and odour.

If unconscious or incontinent a catheter, sheath or pads will be needed (See clinical skills)

Washing and Dressing

Patient may require a bowl to wash or may need to be washed if incapacitated. Wipes, towels and soap are required.

Environment

The patient will require an adequate shelter and to be kept warm.

Exercise / mobilisation

Although rest is often a part of the healing process, exercise is important to your health, it helps you use oxygen more effectively by strengthening muscles which control breathing. Prevents unwanted weight gain and alleviates depression and aids sleep.

If unable to move the patient will need help with passive exercises and rolled to prevent pressure areas developing.

Working and playing, Expressing sexuality, Death and dying and Sleeping

The last Activities of Daily living are more difficult to quantify but must be included in any holistic plan to maintain Activities of Daily living.

Chapter 30 Tactical Considerations

Most of the material in this book can be used in a tactical situation as its designed to be used in remote situations with minimal resources. However military medicine does have some unique factors which are described below.

Whereas in the past explosive trauma generally killed its victim outright the use of modern body armour now provides better protection to the head, chest and abdomen, making wounds more survivable. The upshot of this is that extremity wounds are often severe but survivable.

Initial care in an operational situation is provided if possible by the soldier themselves or other members of the squad all will have basic first aid skills, called 'Care under fire", secondary care is by the team medic who have additional training. The next level is by Combat Medical Technicians (CMTs) who are the military version of Civilian EMTs and paramedics. The final in theatre care given by Nurses and Doctors based at a field Hospital. Current military procedures aims to get the casualty to a field hospital within two hours from time of injury in the meantime the casualty needs to be stabilised and treated

The civilian ABC approach has been replaced with <C>ABC where the first priority is given to stopping catastrophic haemorrhage. In these circumstances there is little point in securing a person's airway or supporting breathing if they are going to bleed out whilst you are doing it.

Military medics use haemostatic agents, chest seals, field dressings and tourniquets to control bleeding. Their use is discussed in detail in the trauma section under wound management.

Use of simple adjuncts such as airways oral and nasal airways are used in both settings, but in a tactical situation surgical cricothyroidotomy are used in preference to intubation. This is partly to do with the technique and the use of a lighted laryngoscope in a potentially hostile situation.

In civilian trauma the mechanism is usually by blunt force which has a higher probability of damage to the neck (C-spine), in military situation where the mechanism is usually through penetrating trauma the likelihood is much less. Therefore the normal procedure of immobilising a casualty is left to the hospital staff to decide this enables the casualty to be moved faster and more easily away from the point of injury. The use of a spinal collar alone doesn't provide complete immobilisation but may be considered as an aid, if a purpose made collar is unavailable it can be fashioned from a sam splint.

For the delivery of drugs and fluids the intraosseous (IO) route is preferred to IV access as the circulatory systems of military casualty ais often impaired due to shock and multiple limb injuries., these routes are discussed elsewhere.

Chapter 31 Triage

The word Triage is derived from the French 'to sort', it has been used in both military and disaster situations to prioritise the treatment of casualties. If the casualty is in cardiac arrest or severely injured and not expected to survive they may be left in favour of casualties that would benefit more from medical attention and evacuation.

In most triage systems casualties are sorted into categories based on an initial survey of their Airway, Breathing and Circulation. When dealing with a mass casualty incident the triage officer will give cards to each casualty showing their Triage Category, there level of consciousness, observations and treatment can be recorded on it. If there condition changes then their Triage Category may also change. The Cards can be folded in different ways to show the various categories.

The initial Sieve is designed to quickly establish priorities a more detailed sort can then take place when resources are available.

The adult Triage Sieve is one method of sorting casualties into one of four categories exist;

Category	Priority	Colour
Delayed	3	Green
Urgent	2	Yellow
Immediate	1	Red
Dead		

Dead

Casualty when checked is not breathing even after opening there airway.

Immediate

Unable to walk & (Respiratory Rate <10 or >29 and/or Central Capillary Refill > 2 seconds)

Urgent

Unable to walk & (Respiratory Rate >9 or <30 and Central Capillary Refill <=2 seconds)

Delayed

Walking Wounded

Triage Sort

The triage sort takes longer to complete as it requires further observations it can change a casualties priority so should be rechecked frequently as resources are available. Total points scored for Glasgow Coma Scale, Respiratory Rate and Systolic Blood Pressure as below.

Glasgow Coma Scale

13-15	4
9-12	3
6-8	2
4-5	1
3	0

Respiratory Rate

10-29	4
>29	3
6-9	2
1-5	1
0	0

Systolic Blood Pressure

90+	4
76-89	3
50-75	2
1-49	1
0	0

12 = Priority 3   Delayed

11= Priority 2   Urgent

<11 = Priority 1   Immediate

0   Dead

Appendix 1 GCS

Glasgow Coma Scale for Adults

Glasgow coma scale 4+	
Response	Score
Eye Opening	
Spontaneously	4
To verbal stimuli	3
To pain	2
No response to pain	1
Best Motor Response	
Obeys verbal command	6
Localises to pain	5
Withdraws from pain	4
Abnormal flexion to pain	3
Abnormal extension to pain	2
No response to pain	1
Best Verbal Response	
Orientated and converses	5
Disorientated and converse	4
Inappropriate words	3
Incomprehensible sounds	2
No response to pain	1

Glasgow Coma Scale for Children <4 Years

Glasgow coma scale (<4 years)	
Response	Score
Eye Opening	
Spontaneously	4
To verbal stimuli	3
To pain	2
No response to pain	1
Best Motor Response	
Spontaneous or obeys verbal command	6
Localises to pain or withdraws to touch	5
Withdrawn from pain	4
Abnormal flexion to pain	3
Abnormal extension to pain	2
No response to pain	1
Best Verbal Response	
Alert, babbles, coos, words to usual ability	5
Less than usual words, spontaneous irritable cry	4
Less than usual words spontaneous cry	3
Moans to pain	2
No response to pain	1

Appendix 2 Medical Terminology

Although You may never hopefully need to know what all those long medical words mean. It is useful to have an understanding of the basics. It means you can converse with medics and if you intend to read more on medical subjects you will know some of the terminology already.

Like most words medical terminology are split into two or more parts, common suffix and prefix's are given below.

a-	Absence of
-ab	Away From
-ad	Towards
Ante-	In Front off
Arthro-	Joint
Brady-	Slow
Cardio-	Heart
Cholecyst	Gall Bladder
Crani-	Skull
Dys-	dysfunction, disorder, difficult or painful
Gastr(o)	Stomach
Glycol-	sugar
Heme-	Blood
Hemato-	Blood
Hemi-	Half
Hepat-	Liver
Hyper-	High, Above, Excess

Hypo-	Low
My(o)-	Muscle
Nephr(o)	Kidney
Neuro-	Nerve
Onco-	Tumour
Osseo	Bony
Oste(o)	Bone
Pneumo-	Lung
Thorac-	Rib Cage
-algia	Pain
-ectomy	Removed by surgery
-emia	to do with blood
-itis	inflammation
-oma	Tumour or swelling
-pathy	Disease
-pepsia	Digestion
-plegia	paralysis
-pnoea	breathing
-tension	Pressure
-uria	Urine

i.e. A-pnoea is no breathing

Dys-pnoea) is difficulty breathing

Tachy- pnoea is fast breathing

Hypo-glyca-emia is Low sugar in Blood

INDEX

12 Lead ECG, 353, 356

3 Lead ECG, 353

4 Cs, 121

abbreviated mental test score, 50

ABCDEFGHI, 13

abdomen, 32, 46, 55, 65, 121, 178, 179, 203, 204

Abdominal Assessment, 45

Abdominal Infections, 273

Abdominal Region, 49

Abdominal Sepsis, 273

Abraisions, 148

Absences, 210

Acarbose, 226

accessory muscles, 19

Acetazolamide, 59, 294

Acne, 274, 275

Activated charcoal, 241

Activities of daily living, 406

Acute Mountain Sickness, 58, 59, 294

adrenaline, 166, 167, 188, 189, 190, 195, 290, 351

Advanced Combat Sponges, 132

Advanced Life Support, 350

AED, 346, 347, 353

Airway, 12, 16, 28, 32, 161, 187, 211, 373

Alcohol, 17

Alcohol consumption, 404

allergic reaction, 70, 187, 188

allergies, 26, 33, 34, 294, 385

Allopurinol, 186

Alpha 2 adrenergic drugs, 280

ALS, 350

amatoxin, 240

Amethocaine, 166, 263, 264, 265

ametop, 166

amitriptyline, 74

Amoxicillin, 196, 270

Ampicillin, 196

Anaerobic/Dental, 275
Anaesthesia, 244
analgesia, 64, 78, 102, 105, 110, 112, 163, 164, 184, 185, 203, 385
Anaphylaxis, 62, 187, 245, 289, 344
aneurism, 77
Angina, 220
angio-odema, 189
ankle, 82, 84, 105
Anthrax, 231, 270, 271, 272, 274
Anti-Anginal Drugs, 287
antibiotics, 117, 142, 148, 149, 151, 176, 187, 198, 204, 383
Anti-Coagulant Drugs, 286
antidepressants, 280
antidote, 65, 161, 276, 277, 352
antihistamines, 69, 70, 187, 188, 294
Anti-Hypertensive drugs, 287
Anti-platelet, 286
anti-venom, 64, 65, 67
Ants, 63

anxiety, 65, 67, 280
Aorta, 179
Aortic dissection, 177
Apendicectomy, 203
APGAR, 399
Appendicitis, 71, 202
appendix, 47
ARDs, 199
Arteries, 122, 123
arthritis, 184, 185, 186, 280
asherman chest seal, 174
aspiration, 335
Aspiration Pneumonia, 197
Aspirin, 219
Asthma, 191, 289, 290
Atenolol, 220
Atrial Fibrillation, 359
Atrovent, 194
Atypical Pneumonia, 196
Augmentin, 148
Auscultation, 30, 43
Auscultation of Heart, 45
Auscultation of Lungs, 43
auto injector, 188

automatic external defibrillator. *See* AED

AVPU, 22

Azithromycin, 242, 270

Bacterial Vaginosis, 275

Bactoban, 63

Bag Valve Mask. *See* BVM

bag, valve, mask. *See* BVM

Basic Life Support. *See* BLS

Battle sign, 27

Bed bugs, 72

Bees, 63

Benzylpenicillin, 237, 271

BinaxNOW, 232

Bird Flu, 197

Bites, 62, 63, 65, 148, 272, 273, 366

Black Widow, 65

blisters, 67, 72, 73, 266

blood glucose, 24, 363

blood loss, 20

blood pressure, 20, 22, 24, 32, 34, 66, 67, 172, 173, 175, 177, 188, 189, 225, 244, 287, 397

blood sugar, 226

BLS, 348

body surface area, 161

Body Systems, 37

bolin chest seal, 174

Bone/joint Infection, 272

bowel sounds, 46

box jellyfish, 67

brachial pulse, 21, 130, 341

Bradycardia, 21, 191, 361

Breathing, 12, 18, 32, 97, 407

Breech Presentations, 403

British Thoracic Society Guidelines, 343

Bronchitis, 270, 273, 276

BSA, 161, 162

bubonic plague. *See* Plague

bucastem, 214

Budesonide, 195

Bundle Branch Block, 360

Burns, 160, 244, 366, 367, 380

Burns Dressings, 370

BVM, 345

Caesarean section, 401

Calamine, 69, 72

Campylobacter, 272, 274

candidosis, 386

Cannabis, 17

cannulation, 305

Capillaries, 122, 123

Capillary Refill Time, 20, 24

Cardiac Tamponade, 28, 120, 177, 352

Cardio Pulmonary Resuscitation. See CPR

Cardiogenic shock, 245

Cardiovascular and Respiratory Assessment, 41

Cardiovascular System, 37

Care under fire, 409

Carotid pulse, 22

catfish, 66

cavity, 89, 117, 118, 120, 143, 144, 175, 178

Cefotaxime, 237, 271

Cefuroxime, 196

Cellulitis, 266, 271, 272, 273, 274, 275

Celox, 132

Cephalosporin, 203

Cephradine, 148

cerebral spinal fluid. See CSF

cervical spine, 28

chest, 29

chest drain, 175

Chest Infections, 273

chest injuries, 89, 173

chest movement, 18

chest pain, 35, 59, 65, 177, 215

chest rise, 29

chest wall, 29

chitosan, 133

Chlamydia, 270, 272, 274, 393

Chloramphenicol, 262, 263, 265, 271

chlorhexidine gluconate, 145

Chloroquine, 232, 233

Cholecystitis, 203

Cholera, 231, 297

Ciprofloxacin, 208, 231, 239, 242, 272, 392, 393

circulation, 12, 20, 32, 74, 75, 80, 86, 87, 130, 131, 175, 206, 243

Circulation, 12, 20, 31

citronella, 62

Clarithromycin, 197, 272

clearing the C Spine, 90

climbing, 10

Clindamycin, 231, 273

Clopidogrel, 219

Closed Pelvic Fracture, 79

Clostridium difficile, 275

Clostridium Tetani. *See* Tetanus

CMT, 409

Co-Amoxiclav, 196, 208, 273

Coblation, 259

collapsed lung, 122

collar, 92, 93, 94, 95, 106, 175

Combat application tourniquet, 132

Combat Gauze, 132

Combat Medical Technicians. *See* CMT

common cold, 257

Compeed, 73

Compression, 114, 115, 169

Compression stockings, 221

Concussion, 169

Congestion, 256

Conjunctivitis, 261

constipation, 46, 267

Continuous infusion, 319

COPD, 217, 289, 344, 345

coral, 67

Corneal abrasion, 265

Corticosteroids, 186

CPR, 20, 348

cramping pain, 65

cranberry juice, 208

cranial nerves, 52

cricothyroidotomy, 339, 410

Crotamiton, 68

croup, 195

CRT, 20, 24

CSF, 27

CSM, 31, 75

CVA. *See* Stroke

cyanide, 161

Cyanosis, 31, 60, 64, 175, 199
cyclizine, 218
cystitis, 207
D&V. *See* diarrhoea and vomiting
Danger, 12, 14
DCAPBTLS, 26, 29, 30, 31
Death Cap, 240
debridement, 121, 152
Deep vein thrombosis, 220
DEET, 62
Defibrillation, 20, 346
Deformity, 53, 77
De-gloving, 149
dehydrated, 25, 26, 35, 307
Dehydration, 25, 56, 244
Dental abscess, 270
Dental anaesthetic, 387
Dental Infections, 273
Dental syringes, 387
destroying Angel, 240
Dexamethasone, 59, 60, 195, 294
diabetes, 224
diabetic, 35, 225, 228, 229

diarrhoea, 25, 66, 202, 244, 267, 273, 311
diarrhoea and vomiting, 240
Diathermy, 259
Diazepam, 101, 210, 280, 294
dicobalt edetate, 161
Digestive System, 37, 39
digital block, 64, 66, 168
dirt bikers, 8
dislocation, 75, 76, 101, 102, 105, 106, 108, 109, 110, 112
distended neck veins, 28, 175, 177
Diuretic Drugs, 288
Diuretics, 172
diverticula, 204
Diverticulitis, 204
dogfish, 66
Dorsalis Pedis. *See* Pulse
Doxycycline, 233, 237, 256, 262, 274, 385, 392, 393, 394
drawing up medication., 313
dressing, 73, 124, 125, 126, 127, 128, 130, 133, 135, 137, 143, 146, 147, 148, 149, 153, 379

DRS ABC, 12
DVT. *See* Deep vein thrombosis
Ear Infection, 273
Ear wax, 256
Ear, Nose and Throat, 254
ECG Monitoring, 353
Eclampsia, 397
Ectopic Pregnancy, 77, 395
ejected, 91, 181
elbow, 84, 98, 105, 106, 107, 108, 129, 131, 307
elevation, 115, 130
Elimination, 407
Emergency Child Birth, 397
Emergency Dentistry, 381
EMLA, 166, 281
end of bed', 11
Endocarditis, 271, 275
Endocrine, Lymphatic and Skin, 37, 40
ENT. *See* Ear, Nose and Throat
ENT Diagnostic Set, 374

Entonox, 164, 218, 284, 372, 390
entry wound, 119
epigastric pain, 215
Epiglottises, 271
EpiPen, 189, 190
Episiotomy, 400, 402
Epithelial tissue, 143
Epithelialisation, 136
Erysipelas, 273
Erythromycin, 196, 266, 274, 392, 393
Eurax, 69, 72
events, 14, 26, 33, 35
exit wound, 120
Expiration, 16
Exposure, 23
External rotation of baby, 403
extremities, 31
Eye Infections, 261
Fahrenheit, 23
Falcipartum Malaria, 273
Falciparum malaria. *See* malaria
FAST, 51, 212

feet, 52, 60, 73, 74, 81, 95, 96, 248

femoral. *See* Pulse

femoral artery, 86, 110, 131, 179

Femoral pulse, 22

Femur, 79, 85, 86, 111

Fentanyl, 101, 279

Finasteride, 209

fingers, 17, 21, 32, 57, 64, 68, 76, 81, 92, 98, 103, 109, 130, 147, 167, 184, 199, 368

first aid, 15, 33, 124, 145, 180, 365

First Aid, 13

fistula, 143, 144

flail segment, 89

flamazine, 164

fleas, 70, 237

Flucloxacillin, 142, 148, 164, 266, 275

fluid requirement, 162

Fluorescein, 265

Fluoride treatment, 386

Focal, 210

forearm, 81, 82, 108, 307

foreign objects, 255, 257, 263

fractures, 27, 75, 76, 77, 78, 79, 80, 82, 84, 85, 87, 88, 89, 91, 102, 105, 108, 110, 117, 119, 121, 147, 169, 170, 173, 182

fragmentation, 116, 117

frostbite, 57, 58, 74

Frostnip, 58

Fucidin, 266

Full thickness, 160, 164

fungal infection, 386

Furosemide, 172

gag reflex, 335

gallbladder, 203

Gastro-intestinal Infection, 272

GCS, 22

Genitourinary System, 37, 39

Gentisone, 255

GI Infections, 273

Gingivitis, 276, 386

Giving Sets, 321

Glasgow Coma Scale, 14, 15, 413, 414

Gliclazide, 226

Glucagon, 230

Glucogel, 229

Glyceryl Trinitrate, 218
Gonorrhoea, 271, 272, 392
Gout, 186
Granulation tissue, 137, 141, 142
GTN. *See* Glyceryl Trinitrate
Gunshot wounds, 116
H5N1, 197
haemoglobin, 343
haemorrhage, 28
haemorrhaging, 20
haemostatic agents, 130, 132, 410
hand, 16, 78, 81, 92, 105, 106, 108, 109, 116, 127, 128, 129, 189, 307, 365
HARM, 114, 115
head injury, 91, 169, 170, 181, 249
Headache's, 267
Head-to-Toe, 26
heamothorax, 89, 175
heart attack, 215
Heat exhaustion, 55, 56
Heatstroke, 56
Helmet Removal, 98
HELPER-R, 400

Hemcon, 133
Heparin, 221
herniated, 169, 176, 179
Herpes simplex, 393
high altitude, 58, 59, 60
High altitude cachexia, 61
High Altitude Cerebral Oedema, 58, 60
High Altitude Pulmonary Oedema, 58, 59, 294
high calibre, 116
high velocity, 87, 116, 117
hip joint, 87
History, 26
Hookworm, 71
horse riding injuries, 9
Humerus, 79, 84, 105
Hydrocortisone, 188, 195, 277, 290
hyp*e*rglycaemia, 227
Hypoglycaemia, 228
Hypothermia, 57, 163, 352, 366, 367, 368
Hypovolaemia, 352
Hypoxia, 352

immobilisation, 80, 91, 96, 406

Impetigo, 266, 272, 273, 275

incontinence, 31

Indigestion, 215, 268

Indirect Pressure, 130

Infected wounds, 142

infiltration anaesthetic, 65, 66, 165, 388

Inflammatory Phase, 134

injections and infusions, 310

insects bite, 62

Inspiration, 16

Insulin, 225

Intermediate Life Support, 350

Intermittent infusion, 320

internal injuries, 76

Internal rotation of baby, 403

intoxicated, 14, 35

intraligamentary anaesthetic., 388

Intramuscular Injection, 317

intraosseous, 324, 410

Intraosseous Device, 373

Intravenous Bolus Injection, 320

intravenous drugs, 318

intubation, 29, 161, 176, 338, 410

Iritis, 262

irregularities, 32

Irritable Bowel Syndrome, 205

isoniazid, 239

IV Fluids, 23, 161, 178

jaundiced, 27

jaw, 17, 92, 102, 103, 104, 177

Jellyfish, 67

ketoacidosis, 227

ketones, 227

kidney stones, 208

Knee, 84, 86, 112

kneecap, 85, 112

Koplik, 234

laceration, 146, 169

Laryngeal Mask Airways, 338

Last Eaten, 33

Leg Ulcers, 275

Legionnaires Disease, 275

Levaquin, 394

Level of Consciousness, 14, 22, 24

Lice, 69, 70

Lidocaine, 166, 168

lignocaine, 163, 281, 328, 387

Limb Amputation, 271

Lipid-Lowering Drugs, 286

lizards, 65

LMA, 338

Local Anaesthetic, 65, 66, 165

log roll, 16, 32

long board, 95

long term care, 406

loperamide, 267

Lorazepam, 101

lower leg, 82, 84

Lung Sounds, 44

Lyme Disease, 270, 272, 274

Lysol, 69

Malaria, 232, 233, 274, 378

Malathion, 68, 70

malnourishment, 138

Malodorous wounds, 143

manage the airway, 335

mannitol, 172

mantas, 66

Maturation Phase, 134

McRoberts Manoeuvre, 401

measles, 233

measure blood pressure, 341

Mebendazole, 71

medical, 13

Medical Assessments, 32

Medical Kit, 365

medical terminology, 416

medication, 23, 26, 34, 59, 64, 172, 187, 194, 203, 299, 307, 366, 370

Medication, 33, 269, 294, 306

Mefloquine, 233

Meningitis, 236, 237, 271, 295, 297

Meningococcal Septicaemia, 271

Metaprolol, 220

Metformin, 225

Methocarbamol, 65

Metronidazole, 149, 150, 197, 203, 242, 275, 385, 386, 393

Miscarriage, 394

mite, 68

moleskin, 73

MONA, 218

Morphine, 101, 164, 218, 279, 280, 281

mosquito, 232

mountain bike, 8

movement, 16, 17, 31, 32, 53, 54, 76, 77, 97, 110, 112, 175, 176, 203, 244

MRSA Associated, 273

mumps, 233

Muscle Relaxants, 105, 280

Muscle spasm, 65

Musculoskeletal System, 37

mushrooms, 240

muzzle velocity, 116

myocardial bruising, 355

Myocardial Contusions, 176

Myocardial infarction. *See* Heart Attack

narcan, 101

nasal flaring, 19

nasogastric tube, 328

Nasopharyngeal Airways, 337

Nateglinide, 226

nausea, 55, 59, 64, 65, 169, 172, 202, 218

nebuliser, 194

Neck, 28

necrotic, 121, 141, 149, 164

needle decompression, 175

Nerve blocks, 165, 388

Nervous System, 37, 38

Nervous system assessment, 50

neuropathic, 74, 281, 282

neurotoxins, 64

neutral position, 91, 95, 181, 330

Nifedipine, 60, 287, 294

nitrous oxide. *See* Entonox

Normal Sinus Rhythm, 357

nose bleed, 256

NPA, 337
NSAIDS, 184, 279
Nymphs, 70
obstructed bowel, 46
Oesophagus, 173, 178
OPA, 335
Open Book Pelvic Fracture, 79
open pneumothorax, 174
Opioids, 280
OPQRST, 26, 35
Oral Fluids, 55, 56
Oral Herpes Simplex, 274
Oral Infection, 274, 276
Oral Pharyngeal Airway, 335
orthopaedic, 53, 121
Orthopaedic Assessment, 53
Oselamivir, 198
Osteoarthritis, 185
Osteomyelitis, 273, 275
osteoporosis, 77, 280
Ottis Externa, 255, 275
Ottis Media, 256, 270, 271, 272

oxygen, 24, 58, 59, 60, 115, 123, 140, 161, 163, 172, 173, 174, 176, 177, 178, 188, 194, 196, 199, 200, 220, 289, 373
Oxygen, 25, 373
Oxygen saturations, 24
Oxygen therapy, 218
oxygenation, 160, 175
Packing wounds, 130
Pain, 15, 35, 36, 59, 66, 77, 90, 135, 165, 177, 185, 204, 281, 282, 307, 384, 417
pain management,, 279
pandemic, 197
paralysis, 66, 417
Partial thickness, 160
Past Medical History, 33, 34
pathogens, 66
patient examination, 11
Patient Position, 11
Peak flow, 192
Pear drops, 17
PEEP, 123
Pelvic Inflammatory Disease, 274, 275
pelvic splint, 87

pelvis, 31, 86, 87, 88, 111, 203
Penetrating wounds, 147
pericardiocentesis, 177
Pericardium, 120
Peridontitis, 274
perineum, 120, 402
periodontal ligament, 389
Peritonitis, 46, 206, 207, 272, 273
PERLA, 22, 53
Permethrin, 62, 68, 70
Phrasing Questions, 33
pinworms, 70
Piriton, 188
Placenta Abruption, 396
Placenta Praevia, 396
Plague, 237
plaster of paris, 80
pleura effusion, 238
pleural cavity, 89
pleuritic, 59, 65
Pneumonia, 77, 178, 196, 197, 198, 199, 231, 238, 270, 271, 273, 275, 276, 378
pneumonic plague. *See* Plague

pneumothorax, 28, 29, 89, 120, 174, 352
poisoning, 240
poisonous, 63, 64, 65, 66, 240
Portuguese man of war, 67
Postpartum haemorrhage, 400
precordial thump, 352
Prednisolone, 195
Pre-Eclampsia, 397
Pregnant Patient, 395
pressure, 20
pressure points, 130
Pressure Sores, 275
PRICE, 114
Primary Assessment, 16
Primary Closure, 151
primary intention, 135, 137
Primequinine, 233
Prochlorperazine, 218
Proguanil, 232, 233
Prolapsed cord, 402
Proliferative Phase, 134
Prophylaxis Diphtheria, 275

prostate, 209
Prostatitis, 272, 274, 275
Providone iodine, 146
proximal interphalangeal, 109
Pulmonary contusion, 174
pulmonary embolism, 220, 222
Pulmonary oedema, 67
Pulse, 21, 24, 32
pulse rate, 20
Pulseless Ventricular Tachycardia, 350
punctured lung, 77, 284
pupil reactions, 27
pus, 27, 141, 142, 143
Pyelonephritis, 272
questioning, 33
Quikclot, 132
Quinine, 233
Rabies, 237
racoon eyes, 27
Radial pulse, 20, 22
rebound tenderness, 47
recovery position, 330
Red Flag indicators', 12

Reduce Salt, 405
Reduce weight, 405
Referred Pain, 76
Rehydration, 55, 56, 369
rehydration fluids, 242
Relenza, 198
renal colic, 208
Repaglinide, 226
rescue hammer, 180
Respiration Rate, 18, 24
Respiratory System, 37, 38
Respiratory Tract, 275, 367
Respiratory tract Infection, 272
Respiratory Tract Infection, 270
Respiratory Tract Infection, 272
Response, 12, 14
Retained Placenta, 400
Reteplase, 219
Retinal Haemorrhages, 60
retractions, 19
ribs, 35, 89, 175, 356

Rifampicin, 231, 239
road traffic collisions, 180
Rosacea, 274, 275
Roundworm, 71
RTC, 91, 170
Rubella, 233
Rubin Manoeuvre), 401
Rule of 9s, 161, 162
runners injuries, 9
ruptured spleen, 77
RUQ, 203
Salbutamol, 188, 193, 194, 289, 380
Salmonella, 272
SAM Splint, 80
SAMPLE, 26, 33
SARS, 199
Scabies, 68
Scene Assessment, 14
Sciatica, 279
scorpion fish, 66
Scorpions, 64
sea anemones, 67
sea coral, 67

secondary assessment, 23
Secondary Closure, 151
secondary intention, 136, 137, 145
seizures, 31, 66, 210, 211, 281, 294
sensation, 29, 31, 32, 52, 57, 76, 77, 97, 171, 266
sensitive teeth, 386
Septic Arthritis, 273
Septic shock, 245
Septicaemia, 57, 271, 272
Sexually Transmitted Diseases, 392
Shigellosis, 272
shock, 11, 19, 20, 55, 76, 97, 176, 240, 243, 244, 245, 246, 247, 248, 249, 250, 355, 396
Shortening of the limb, 85
shotgun, 118, 119
shoulder, 52, 77, 78, 84, 95, 96, 105, 106, 107, 128, 129
Shoulder Dystocia, 400
Signs, 14, 33, 34, 59, 60, 191
simple pneumothorax, 174
sinus, 143, 144

Sinus Bradycardia, 358
Sinus Tachycardia, 358
Sinusitis, 270, 273, 274
skiing injuries, 9
Skin / Soft Tissue, 272
skin colour, 24, 25
skin grafting, 149
Skin Infections, 275
Skin Staples, 154, 376
skull fracture, 27
sling, 82, 84, 115, 128
Slough, 141
snakes, 64
Snow blindness, 263
Solvents, 17
Specialist Diagnostic Kits, 378
Spenco second skin, 73
sphygmomanometer, 341
spider, 65
spinal damage, 31
Spinal Immobilisation, 16
spine, 16, 32, 89, 90, 96, 97, 107, 182, 410
sprain, 75, 113

Spyroflex, 73
ST Elevation, 359
Standing Take Down, 95
STD, 392
Steri-strips, 152
steroids, 184, 195, 280
stethoscope, 30, 43, 46, 342
stingrays, 66
Stings, 62, 63, 64, 66
strain, 75, 95, 96, 113, 114
Streptococcal, 275
stroke, 211
sub dermal injection, 315
subcutaneous emphysema, 29, 174
suction, 335
sumatriptan, 214
sunburn, 55
Superficial, 160
suprapubic pressure, 401
Supraventricular Tachycardia, 359
Surgical Prophylaxis, 271, 272
suture, 144, 155, 157, 158
Suture Set, 376

sweating, 55, 66, 67
Swelling, 28, 53, 59, 62, 77, 85
swelling of eyelids, 65
Swine Flu, 197
Symptomatic Relief, 406
Symptoms, 34, 59, 60, 65, 90, 184, 185, 191, 198, 204, 393, 397
Synometrine, 294
Syphilis, 274, 392
Tachycardia, 21, 362
tactical situation, 409
Tamiflu, 198
Tamsulosin, 209
TB, 238
tea tree, 164
Temperature, 24, 32
temporal. *See* Pulse
temporomandibular joint, 102
tenderness, 32, 77, 90, 114, 160, 202
Tenecteplase, 219
Tension Pneumothorax, 28, 174
Tertiary Closure, 151

Testicular infection, 394
Testicular Torsion, 394
Tetanus, 238
Thiazolidinediones, 226
thoracic cavity, 178, 179
Threadworms, 70
three P's, 13
Throat Infections, 271
Thrombolysis, 219
Thrombolytic Drugs, 288
Thrombosis, 352
TIA. *See* Stroke
Ticks, 63
Tonic Clonic, 210
Tonsillectomy, 259
Tonsillitis, 258
Topical anaesthesia, 165
Tourniquet, 20, 130
Toxic Shock Syndrome, 394
Toxins, 64, 242, 352
Tracheal deviation, 28
traction, 86, 106, 107, 108, 109, 111, 112, 406
traction splints, 86
tramadol, 164, 280, 281

Transcutaneous electrical nerve stimulation, 282

Trauma, 13

traumatic injury, 11, 75

Triage, 411

Triage Sieve, 411

triage sort, 412

triangular bandage, 128, 129, 130

Trimethoprim, 208, 276

tripod Position, 19

Tropicamide, 263

Tuberculosis, 238

Turgor, 24, 25

tympanic membrane, 255

Typhoid, 239, 270, 271, 272, 297

Typhus, 70

unconscious, 13

upper arm, 84, 105, 289

Urethritis, 207, 270, 274, 275

Urinary Catheterisation, 325

Urine Dip Sticks, 378

UTI, 207, 208, 272, 274, 276, 393

vaccine, 198, 200, 234, 238, 295, 297

Vaginal Discharge, 393

Veins, 122, 123

Vena Cava, 179

Ventricular Fibrillation, 290, 347, 350, 360

Ventricular Tachycardia, 290, 347, 361

vital signs, 24

vomiting, 25, 59, 65, 66, 202, 244, 311

warfarin, 221

Wasps, 63

waterjel, 163

weakness, 65, 187

Whipworm, 71

Whooping Cough, 200, 275

Wilderness Medical Kit, 379

Wood-Screw Manoeuvre, 401

Wound healing, 134

xylocaine, 166, 167

Zanamivir, 198

Zavanelli Manoeuvre, 401

Printed in Great Britain
by Amazon.co.uk, Ltd.,
Marston Gate.